H. H. Nininger

Master of Meteorites

H H. Nininger

Master of Meteorites

Kenneth J. Zoll

VVAC Press
Camp Verde

VVAC Press Edition, February 2023

Published by VVAC Press, a division of the
Verde Valley Archaeology Center
460 W. Finnie Flat Road, Camp Verde, AZ 86322
928-567-0066
Email: vvacpress@verdevalleyarchaeology.org

International Standard Book Number: 978-0-9820378-9-8

Cover image: istockphoto.com image by Maxvis

For Nancy who,
like Harvey's Addie,
provides unseen love and support

⌘

Proceeds from this publication go to the
Verde Valley Archaeology Center to
further its mission of conservation,
preservation and education.

CONTENTS

PREFACE AND ACKNOWLEDGMENTS

You may be wondering why an archaeology organization is publishing a book about meteorites and Dr. Harvey H. Nininger. The Verde Valley Archaeology Center was formed in 2010 in Camp Verde, Arizona, for the purpose of conservation, preservation and education of the artifacts, sites and stories of the Native Americans who made this area their home for 12,000 years. For research purposes, one of our first tasks was to accumulate every report, paper, article, thesis, dissertation and highway report relating to the archaeology of the area. During this effort, I came across a small article in *Plateau,* the quarterly publication of the Museum of Northern Arizona in Flagstaff, entitled "A Meteorite Discovery with Archaeological Importance from the Camp Verde Area, Central Arizona," by Dee T. Hudson with Arizona State University (Volume 45, Number 1, Summer 1972).

The idea of a meteorite having archaeological importance immediately drew my attention since my specific area of interest and research is archaeoastronomy, or cultural astronomy. From the article, I learned that this meteorite had been in the collection of Dr. Nininger and was currently at the Arizona State University's (ASU) Buseck Center for Meteorite Studies in Tempe, Arizona. The original archaeological location of the meteorite was rather vague and was based on the area as it was in 1935, which had seen significant development and road changes since that time. With a great deal of good luck, I was able to locate a photograph, in the archives of the Denver Museum of Nature and Science, of Dr. Nininger standing in the ruin where the meteorite was found. Matching the surrounding terrain, the location was identified and visited.

To learn more about this meteorite, I visited ASU and Dr. Laurence Garvie, Collection Curator and Research Professor, at the Buseck Center for Meteorite Studies. I was able to review the files and learn more about the Camp Verde Meteorite. I thought that if we were able to acquire the meteorite on loan for our new museum, it would generate a great deal of public interest. Long story short, we were allowed to "borrow" the meteorite in 2015 for nearly a year to a great deal of press coverage and public visitation. This started me on a different path in archaeoastronomy, namely, the study of the uses of other meteorites among ancient cultures. I learned that ASU had two other meteorites, owned by Dr. Nininger, that were found near other sites in the area. All three of these were fragments from the Canyon Diablo meteorite that formed Meteor Crater. This lead to more research and a paper on how they arrived in the Verde Valley, at least 50 miles from Meteor Crater.

In 2021, the VVAC was able to purchase a much larger building with three times the exhibit space. I immediately thought that it would be a great exhibit to have all of the archaeological meteorites on display for our grand opening. Another long story short, thanks to Dr. Garvie and ASU, this became a reality when we opened to the public in early 2022. Many people asked about Dr. Nininger and I was dismayed that almost no area resident had ever heard of him. In fact, I had a visit from a professor from Argentina who was looking for the location of his former Sedona Museum. No one at the Sedona Visitor Center knew of him or the museum. I found this upsetting and decided that needed to change - thus the genesis of this short book.

Putting this book together required several visits to ASU's Hayden Library Archives to review the collection of research notes, correspondence, newspaper clippings, drafts and final versions of books and articles, photographs, and other materials donated by Dr. Nininger. I also reviewed four three-inch binders of material that he had donated to the Sedona Historical Society. I am extremely grateful for the assistance and patience of the staff of the ASU Hayden Library Archives, and Janeen Trevillyan of the Sedona Historical Society. I am especially grateful to Dr. Laurence Garvie who provided insight to Dr. Nininger and his collection, as well as access to many of the collection items, including 57 of Dr. Nininger's notebooks on his explorations.

Very special thanks goes to Jim Bank, Gary Huss, and Peggy Schaller, grandchildren of Dr. Nininger, for their assistance, review and suggestions.

Ken Zoll
Sedona, Arizona

INTRODUCTION

> "he has a lot of what Sherlock Holmes had. He follows clues and scents with uncanny skill."
>
> "the world's foremost meteorite sleuth and miner"
>
> "America's No. 1 meteorite-sleuth until one hits him on the head."
>
> "the world's most persistent star-dust chaser"
>
> "the indefatigable collector and student of meteorites"

The above are some of the characterizations of H. H. Nininger by various writers of articles about him over his career. As I researched the various magazine and newspaper articles, covered in the first chapter, about H. H. Nininger, it struck me that he was, in a sense, a master at manipulating the social media of the day. Before the Internet there were magazines and newspapers that reached into almost every home in America. He was a prolific writer and lecturer about his passion and he spread his knowledge and passion widely. As you will see, this greatly benefited his search for the holy grails of the universe on earth.

The fruit of these efforts was the development, as described in Chapter 2, of what would eventually become the largest collection of meteorites in the world! This was accomplished even after he was told by one of the most respected members of the scientific community of the day that he was wasting his time — he would be lucky to find just one meteorite in his lifetime. One writer described him as a "courteous, country-boy, straight-forward midwestern" individual. His success was attributed to "a rare ability to communicate his knowledge and his enthusiasm, with an engaging rustic simplicity of approach." To use his own words, he was "backed up by a strength of purpose" that was unmatched at the time.

All of these personality traits came to play in Chapter 3 that describes the "strength of purpose" of his study of Meteor Crater. His brilliant, inventive, and inquisitive nature faced obstacles along with skepticism, and sometimes hostility, from "associates" in the field. But he never lost sight of his purpose, and eventually achieved the recognition he deserved.

As he aged and work in the field became more difficult, he and his wife Addie, sought to show their collection to the public through the establishment of a museum. Chapter 4 reviews their museums, the first located near Meteor Crater, and then moving to Sedona when Route 66 lost traffic to the new Interstate highway.

The Niningers eventually retired and closed their museum. They sold their collection to the British Museum of Natural History and to Arizona State University. This is covered in Chapter 5 along with a description of his "post museum" life.

If you are unfamiliar with the study of meteorites, an Appendix is provided that may be helpful. However, the intent of this book is not to provide a course in meteoritics, but rather to tell the story of a man consumed with discovery and exploration of an relatively ignored branch of earth and space science.

CHAPTER 1

A Comet Strikes the Earth

Leave the beaten track occasionally and drive into the woods. Every time you do so you will be certain to find something that you have never seen before.

—- Alexander Graham Bell

Educational Awakening

Harvey H. Nininger was born on January 17, 1887, in a one-room house about four miles southwest of Conway Springs, Kansas, in Sumner County. He lived there with his family until 1895, when he was eight years old. Because of the severe wind storms in Kansas at the time, his father decided to move to "a less frontier neighborhood." They moved to the foothills of the Ozark Mountains near Warrenburg, Missouri. "Our house was a far cry from that we had left in Kansas."[1]

They remained there for four years until his father decided in 1899 to move to Oklahoma midway between Guthrie and Stillwater. His two older brothers drove a covered wagon on a covered wagon train that was going from Missouri to Oklahoma. His father rode a stock train with their horses and cattle. Harvey, with his mother, younger brother and newborn baby sister, followed by passenger train.

Public education, as it is known today in the United States, originated in the 1830s with the American Common School Movement. Common schools institutionalized the idea of free, public, nonsectarian or "common schools" for children. Bringing together pupils of all cultural and ethnic backgrounds, the mission of the common school was to teach the ideas of democracy, civic responsibility, and Americanization to children and their parents, especially if they were immigrants.

Common schools were formally developed in Oklahoma after the Oklahoma Constitution of 1907 mandated a free, public education for all children. Harvey attended the equivalent of a common school for a few years. It was a "very hit-and-miss" type of schooling.[1] School would open in September and he would attend between the times he was needed for picking cotton, which would end in November. He would then return to school until March when the next planting took place.

His parents did not think much of education beyond the common school and he admitted that he did not think much beyond that either. However, that changed in the summer of 1907 when he did some work at a local orphans' home to earn some money. While there he met some people away from home for the first time. During the 1976 Flagstaff (AZ) Oral History Project interview, he stated that after he met these people "I very definitely decided I needed an education (chuckles)." At age 20, Harvey decided it was time to move out and get an education.

In 1896, a bill to establish a normal school at Alva was introduced and passed in the territorial legislature. This established the Northwestern Territorial Normal School at Alva, the second such school in the Oklahoma Territory. Harvey's brother was living near Alva and wanted to go to college there so he invited Harvey to come and live with him. They attended together for two years. He graduated with a major in biology which stimulated his interest in science.

Figure 1.1. Harvey Harlow Nininger. Courtesy of Sedona Historical Society.

Nininger wanted to pursue his education within any science. His parents were afraid that his religion "would be spoiled if I got further into science." They wanted him to go to a church school. So he returned to Kansas to attend McPherson College, founded in 1887, as a four-year liberal arts college related to the Church of the Brethren. In his third year, he was offered a summer job working in the museum. The director of the museum liked Harvey and thought much of his abilities. The professor was going to Harvard in the fall and he asked Nininger to take over his biology classes. Harvey initially refused since he was only a junior, but the president of the college persuaded him to take the job. Nininger later recounted, in the oral history interview, that this was an extremely valuable chapter in his life because when he graduated in 1914 at age 27, he graduated with teaching experience.

Figure 1.2. Wedding photo of Harvey Nininger and Addie Delp. Courtesy of Sedona Historical Society.

Harvey married Nancy Adeline "Addie" Delp (1892-1978) on June 5, 1914. She would be his partner in life and in the pursuit of meteorites. After gradation, Harvey took a job as head of the biology department at La Verne College, about 35 miles east of Los Angeles, California. While teaching he earned a master's degree in Biology from nearby Pomona College in Claremont, California in 1917. In 1918, with World War I in progress, he took a position as field representative in South Dakota with the Department of Entomology, a unit within the Federal government. That position, however, only lasted to the end of the war in November 1918.

Figure 1.3. Vintage postcard of McPherson College, circa 1900.

Nininger was offered a position with Southwestern College in Winfield, Kansas, doing field insect control throughout Kansas, and teaching in the biology department. But in 1920, he was offered a position at McPherson College. His former professor was retiring and they presented an offer he could not refuse. This was a fateful decision that would lead to a total change in the direction of his life.

1923 Epiphany

Nininger would soon experience a series of events that would focus his "naturalist" interests onto meteorites. First, while attending the 40th annual meeting of the American Ornithologists' Union in Chicago in October 1922, he took the opportunity to visit the famous Field Museum's mineralogical hall. Accompanied by a "noted scientist from a Philadelphia museum," they came upon the meteorite exhibit. Nininger admitted that the word "meteorite" was rarely mentioned in any of his college classes and, when it was, it was dismissed as "irrelevant." His associate commented: "I wonder if they really know that these things come to earth from space." Nininger said that he had "spoken my own thoughts."

As a naturalist and biology professor, Nininger regularly read *Scientific Monthly*. In the November 1923 issue,[2] Nininger was captivated by an article by Professor Arthur M. Miller, University of Kentucky, simply entitled "Meteorites." The article caught his interest from the first sentence:

> "Meteorites, being solid bodies which come to earth from celestial space heated to dazzling brilliance on the surface from friction with the air and detonating with terrifying intensity as they set up violent atmospheric compression and rarefaction waves, would appear to carry with them proofs of their extramundane origin as incontestable as could well be demanded."

Figure 1.4 Coon Butte Crater from photograph courtesy of Dr. George P. Merrill as it appeared in "Meteorites" by Arthur Miller, *The Scientific Monthly*, Vol. 17, No. 5 (Nov., 1923), 440.

The article reads like a primer on the study of meteorites, describing iron, stony and "intermediate" meteorites and their composition. Perhaps it was fate or destiny that brought this article to Nininger's attention, for it included a photo and description of the "Coon Butte Crater, Arizona" (later renamed Meteor Crater) where he would spend many years exploring and studying (Chapter 3).

In his oral history interview, Nininger said that prior to visiting the Field Museum and reading Miller's article, he had never heard anything about meteorites. Shortly after reading the article he told his wife "I'm going to make a hobby of this. It's just too interesting to pass up."[3]

Nininger recounted what he "would not give for the opportunity to witness the fall of a meteorite to earth!" Perhaps the article was on his mind when, on a chilly evening, Nininger was standing outside with a fellow professor at McPherson College on November 9, 1923. "Suddenly a blazing stream of fire pierced the sky, lighting the landscape as though Nature had pressed a giant electric switch," Nininger recalled. "The blade of light vanished . . . leaving a darkness seeming thicker than before."

Both men were left speechless for a time. After seeing the fireball, Nininger said that he was "all cocked and primed and started right in to chase that fireball." He began to sketch its probable path and declared "I'm going to hunt for it." He then developed a procedure to locate meteorites that he would follow for many years. First, he knew that others must have witnessed the fall so he wrote to the major newspapers in Kansas and asked for the public's help with the following information:

1) What was the person's location when they made the observation?
2) What was the direction of the fall from their location ?
3) How high above the horizon was the fireball when it disappeared or was extinguished?
4) Was any sound heard with, or shortly after passing, the fireball?

Reports came flooding in. This was the first method developed for locating witnessed meteorite falls. He would collect eyewitness reports, compass readings, elevation data and the trajectory of the path through the atmosphere. This would allow him to pinpoint the end point of the phenomenon. From these data a probable strewn-field area for the landing of the meteorite would be determined.

Applying this technique to the meteorite he had just witnessed, he sifted through the most helpful information, and combined it with his own observation. He estimated that it had landed somewhere in the vicinity of Coldwater and Greensburg, Kansas. This was about a six-square-mile area. Despite the slender chance of finding it, he could not resist trying. He began to search the area on weekends with small volunteer groups of students.

He knew that most meteorite falls were not witnessed, especially those in the distant past. To recover meteorites from unwitnessed falls, he developed a program of educating the general public, and particularly rural populations, to the existence of meteorites. This entailed the publication of articles in magazines and newspapers. He would offer lectures in public schools from grade school through high shool, to service and farm organizations, and do house-to-house canvassing in rural areas. As an incentive for participating in the program, he would make an offer to purchase all meteorites, generally at $1 per pound.

1925 Family Trip East

In the unpublished, extended version of his autobiography, *Find a Falling Star,* Nininger confesses that for "several years I had been noticing in my own life a slow but appreciable decrease in the quality which enables a man to attack the difficult and to try the untried. Our civilization softens and tones down too much those attributes of primitive man which enabled him to cope with problems of life in the wilderness."[4] In December 1924, he was reading the January 1924 edition of *National Geographic Magazine* that contained an article on explorations of Carlsbad Caverns in New Mexico. After reading the article he told Addie "We're going to see those wonderful caverns next year!" Addie was a little surprised since she was pregnant and expecting to deliver in March. Harvey responded to her concerns by telling her "We'll take the whole year off and travel. The baby can celebrate its first birthday on the road."

The practical Addie asked how they would finance such a trip. Harvey indicated that he would lecture in schools as they went. He figured that they could live cheaply on the road, and if he could get only ten lectures a month at $25 each it would equal his salary. "Besides, we may find salable fossils or meteorites. If worse comes to worse, I'll pick cotton. I can still do some of the jobs that got me through school."

Fortunately for Harvey, Addie loved to travel. They decided to travel when Addie gave birth to their third child, Margaret, and began adjusting their ideas of traveling with an infant.

The "House Car" was built before the days when motor homes were available commercially. Many campers produced in the early days were built around plans supplied to a carpenter or cabinet maker. The Niningers had seen magazine pictures of a few vehicles fitted with living quarters and had examined one example. They added ideas of their own to the features shown in the magazines. They employed a German carpenter who believed in sturdy construction. The "house" would rest on two sixteen-foot oak two-by-tens, and every joint would be mortised solidly and held together by screws or bolts.

At the front of the vehicle was a clothes closet and writing desk, furnished with compartments above and shelves below, for reference books and a typewriter. Each side of the house-car opened out in wings, and cushioned seats unfolded into two full beds. While traveling, the bedding was strapped to the walls above the seats. The baby's bed was on wheels and could be moved from one end of the house-car to the other. A small rocking chair could ride either in back or next to the driver's seat.

The kitchen had a sink with running water (carrying sixteen gallons), a gas stove, work table, drawers and cupboards and an ice-box. A trap-door in the floor opened to storage for vegetables and canned goods. A second trap-door was for laundry. Out the back door were lockers which opened from the outside.

Figure 1.5. Nininger house-car. Courtesy Sedona Historical Society.

Once the "house" had been constructed, the next decision was to determine upon what to mount it. Since their itinerary was to go into thinly settled, or uninhabited country, the car had to be as dependable as possible. They knew that Ford parts could be purchased at all garages and even at ten-cent stores and soda fountains. So for $900 and their old car, they acquired a new Ford Model-T truck chassis, and for another $900, they contracted to have their "house" mounted on it (Figure 1.5).

When the vehicle was completed, Nininger took leave from his college work for the first of the family trips. They took off in their "house-car" on a journey across half the country (Figure 1.6). Addie began her diary of the "Nininger Runabout" with a notation that the house proved to be neither fly-proof—though its windows were screened—nor waterproof. One night, she noted, the family slept between rows of pans.

Addie recounted that "At every stop our house was surrounded by people; some admirers; some scoffers; some merely curious; and some skeptical. . . . Some of the things that people said to me when we were getting ready for the trip were enough to discourage a less determined or adventurous spirit but it only made me more anxious to show them it could be done and successfully done in every way."[5]

Figure 1.6. Nininger house car on the road. Courtesy Sedona Historical Society.

The lecture field offered possibilities of income along the trip. Nininger's natural history presentations, bolstered by "magic lantern" slides, had proven popular. The meteorite talks that he had given in connection with the Coldwater search had been received "with more than ordinary interest." The model of magic lantern that Nininger used is unknown but most magic lanterns used a single-lens, utilizing two or three wick oil burners (Figure 1.7). These lanterns were popular with traveling showmen, fraternal lodges, schools, and in the home. Oil burners were popular with traveling shows, especially in rural areas where neither electricity nor adequate supplies of gases were easily available for the production of light.

Figure 1.7. Magic lanterns came in many shapes, sizes, and configuration. This is a typical traveling showman variety.

They left McPherson in September 1925 with the intention of making a circuitous tour of the Southwest, ending up in Palmer Lake, Colorado for the annual Rocky Mountain Summer School in May 1926. About 120 miles along their trip was the town of Paradise, Kansas. Nininger was able to book four lectures in the Paradise vicinity. One of the talks provided a bonus — nearly five years after a talk, a boy who had attended one of them, contacted Nininger about a meteorite. For Nininger, this furnished proof that his meteorite recovery method was correct with the first meteorite that he could trace directly to his educational efforts.

In southern Kansas, he obtained a twenty-pound meteorite specimen of the Brenham fall (Chapter 2). He immediately sent it to the National Museum in Washington, D.C., for cutting. He would give them half of the meteorite for the cutting cost and they would return the remainder in the form of salable slices. This would be a resource for ready cash should it be needed somewhere along the line. They had packed several nice meteorite specimens from home to help illustrate lectures, but they could be sold if needed.

In Austin, Texas, he gave six lectures that averaged $9. Corpus Christi turned out to be a much better financial stop. Nininger sold three meteorite specimens to the school board chairman for $270. He was then told that arrangements would be made for seven lectures to be given the following week. The lecture series did not meet with great demand in other Texas towns, although at McAllen one lecture netted thirty dollars on the ten-cent admission plan.

They traveled to El Paso and then on to Carlsbad Caverns (the inspiration for the trip), and to other towns in New Mexico. In Arizona they visited Benson, Tucson, Casa Grande, Phoenix, and then on to Prescott, Jerome and the Grand Canyon.

As they passed Flagstaff they began to watch for the first sign of one of the chief goals of the trip, "the great hole in the earth dug long before memory by a tremendous meteorite." They identified the broken-rimmed crust rising above the level plain of the desert, watching the elevation take on the appearance of a low cone, "a cone that differed from ordinary volcanic peaks in its being light-colored and composed of fragmental sediments rather than the usual volcanic products."

They drove to the base of the slope leading to the rim of the crater, where they pitched camp for the night. As Nininger recounted:

> "No other spot on the earth's surface could furnish more inspiration for imaginative conjecture than the crater, this spot where a belated planetesimal arrived sufficiently late in the process of earth-building to leave for man's contemplation the stone-shattered entrance to its mountainous burrow."

After a night on the wall of this "memorial to a cataclysm," they studied the pit, the rim, and the surrounding plain by daylight. He did not know then that in a few years, he would live alongside this crater.

They passed through the Painted Desert and Petrified Forest. Nininger examined a large meteorite on display at Navajo Station, and then traveled east through New Mexico. He purchased a small meteorite in Albuquerque and drove on to Santa Fe. They continued on their journey to Palmer Lake, Colorado, for the start of the annual Rocky Mountain Summer School in June 1926.

The Rocky Mountain Summer School (Figure 1.8) program flyer stated: "Many teachers and students are annually compelled to suspend their educational activities during the intense heat of the summer on the plains, and seek recreation at the expense of education in the mountains. A natural need has consequently given further justification for the establishment of the Rocky Mountain Summer School, where standard educational work may be continued while enjoying the recreative surroundings, which the mountains alone an afford."

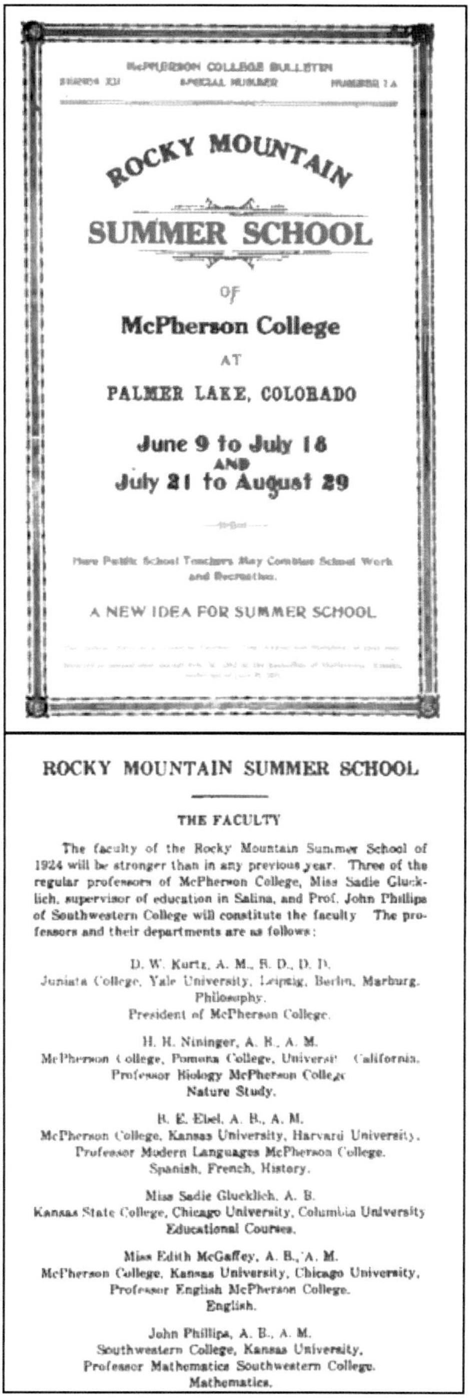

McPHERSON COLLEGE BULLETIN

SPECIAL NUMBER

ROCKY MOUNTAIN

SUMMER SCHOOL

OF

McPherson College

AT

PALMER LAKE, COLORADO

June 9 to July 18

AND

July 21 to August 29

Where Public School Teachers May Combine School Work and Recreation.

A NEW IDEA FOR SUMMER SCHOOL

ROCKY MOUNTAIN SUMMER SCHOOL

THE FACULTY

The faculty of the Rocky Mountain Summer School of 1924 will be stronger than in any previous year. Three of the regular professors of McPherson College, Miss Sadie Glucklich, supervisor of education in Salina, and Prof. John Phillips of Southwestern College will constitute the faculty. The professors and their departments are as follows:

D. W. Kurtz, A. M., B. D., D. D.
Juniata College, Yale University, Leipzig, Berlin, Marburg.
Philosophy.
President of McPherson College.

H. H. Nininger, A. B., A. M.
McPherson College, Pomona College, University California.
Professor Biology McPherson College
Nature Study.

R. E. Ebel, A. B., A. M.
McPherson College, Kansas University, Harvard University.
Professor Modern Languages McPherson College.
Spanish, French, History.

Miss Sadie Glucklich, A. B.
Kansas State College, Chicago University, Columbia University
Educational Courses.

Miss Edith McGaffey, A. B., A. M.
McPherson College, Kansas University, Chicago University.
Professor English McPherson College.
English.

John Phillips, A. B., A. M.
Southwestern College, Kansas University.
Professor Mathematics Southwestern College.
Mathematics.

Figure 1.8. Pages from the 1924 program for the McPherson College Rocky Mountain Summer School. McPherson College Archives.

When they returned to McPherson, they had been away nearly a full year. The children went back to catch up with their schoolmates and Addie and Harvey returned to their home and college chores, "but we had tasted roving independence and wished for more of it." He continued to have hopes of somehow making a livelihood out of his interest in meteorites.

No new meteorites had come to light as a result of his lectures during the house-car tour of Kansas, Oklahoma, Texas, New Mexico and Arizona, but he had been able to buy a few specimens. One, the small Ballinger iron purchased in a mineral store in Manitou Springs, Colorado, at the end of their trip, proved to be a new fall.

As they approached the close of the house-car trip, Nininger noted that "the year's travel had done something for us that was in the nature of a rejuvenation. I had regained the zest and dare that had characterized my life after breaking away from cotton farming and embarking on an educational career. And Addie, in spite of the tremendous workload created by travel with three small children, felt that she wanted more of this kind of life."

1928 Student Trip

Nininger noted that there was no recognized science of meteoritics. Dr. Farrington had suggested the term *meteoritics* as the name for the study of meteorites. Those who were working with meteorites at the time were mineralogists, astronomers, chemists, geologists, who only performed meteoritical studies as side projects. The leading students of meteorites had told Nininger that a science of meteorites was a "wild dream," and that the foundation for such a science would have to be a plan for greatly multiplying the quantity of meteorites available for study.

Nininger learned what he could on his own, collected meteorite specimens from recorded meteorites falls, and continued to lecture on meteorites. The visible result of his lectures was a keen interest shown by the audiences. His immediate problem, however, was to add to his store of marketable material while continuing in the employ of the college. The College continued to make it possible for him to carry on with the meteorite program by keeping his teaching load light.

Nininger thought that perhaps another year of leave from his regular teaching, a year devoted to a student tour, would provide an opportunity to learn more about meteorites and to test his theories for their recovery. He thought that the literature on the subject of meteorites was inadequate. He decided he would need to undertake another extended trip to examine the major meteorite collections in Washington, D.C., Chicago, New York, Harvard, Yale and Amherst. He developed a plan for a long-term, large-scale recovery of meteorites, while taking college students on a tour to some of the sources of the natural histories they studied in class.

After a year on campus, he and Addie were ready to undertake a natural history trek with a class of students who would be working for college credit. Fortunately, the McPherson College board of trustees accepted his proposed program. He was granted a traveling professorship. He volunteered to forego a salary and instead collected a $75 tuition fee from each of the students.

The trek was to cover the school year of 1927–1928. The students would work the entire school year for eighteen hours of credit. As applications for membership in this unusual class arrived, he scheduled meetings at regular intervals to brief prospective students and prepare them for what he knew would be a difficult year. Students were to furnish their own transportation, camp equipment and sustenance. The program would include lectures, and students would keep notebooks that outlined field work. Credits could be earned in zoology, botany and geology.

Nininger had grown impatient with the tendency to make college life too easy on the students. He saw the practice of feeding students their education in lecture form, rather than having them find out things for themselves, as largely an exercise of memory. His program would challenge the students' ingenuity, powers of observation, thought and endurance. Thirteen students comprised the class. The Rocky Mountain Summer School term was to close the first week in August. The group would gather at Palmer Lake and depart on the Natural History Trek on August 22.

They were organized into three student groups. The Nininger family was the fourth group in their own car. Each group purchased a new two-door Pontiac sedan equipped with a large trunk. An additional cabinet was built along the left side of each car. One car carried five men, another four, the third carried two coeds and a married couple. Each student had a roll of bedding with a suitcase or two. Each car carried a tent, stove and cooking utensils, table and chairs, and a small basic library of science texts. Each group cooked and ate as a unit. Rules were adopted for the purpose of keeping together in travel, and all were to set up camp within calling distance.

Their first stop was Yellowstone National Park where they studied the geology of the area, including its volcanics, petrified forest, hot springs,

Figure 1.9. Cars ready to begin the Natural History Trek. Courtesy of the American Meteorite Laboratory Photo Archives, Collections Research for Museums, Denver, CO.

geysers and boiling mud pots. They cataloged the plant life, birds and mammals. From Yellowstone they went into Idaho and south to Salt Lake where they studied the water birds on the salt marshes. Nininger had arranged with the National Audubon Society to make observations of bird life throughout the trip. The trek continued on to Bryce Canyon, Zion and the Grand Canyon national parks. Continuing on, they stopped in St. George, Utah, Las Vegas and eventually Los Angeles. They then turned east to spots from the earlier family trip including Jerome, Montezuma Castle, and the Camp Verde Salt Mine.

They visited the Cornville area where they found fossil foot prints. Dr. Byron Cummings of the University of Arizona in Tucson has agreed to purchase any such finds. When they arrived in Tucson they were rewarded with a check for $380 for the fossil prints.

Nininger made a two-day side trip to the noted copper mine at Cananea, Mexico, accompanied by one of the students, Alex Richards, who was fluent in Spanish. They made contact with a local resident who owned a sixty-eight pound meteorite. He was able to buy this specimen of the **Arispe** meteorite. This was an iron meteorite that had been used as an anvil for several decades but seemed little the worse for all the hammering it had received.

Since the start of the trip, Nininger brought a forty-pound Canyon Diablo meteorite from Meteor crater. He received an order for a slice of that meteorite, so while in El Paso, Texas he went into a large machine shop to inquire about having a piece cut off that would measure four to five square inches. The job took several hours and several saw blades. The superintendent refused to let him pay for anything more than the cost of the blades, but he asked a good many questions about meteorites.

They continued on to New Orleans and then to Key Largo, Florida, studying the flora and fauna at these diverse climates. Turning north they went through the Okefenokee Swamp in Georgia. They continued travelling north, colleges, universities, museums, libraries, research establishments, historic points and industrial plants were now the subjects of their studies. During the following eight weeks they visited the most important educational centers of the East.

Nininger would visit likely hills or farms where people said they had objects like the one he was lecturing about. He found dozens of meteorites this way and would sell some to local museums to keep the family eating well.

He kept a notebook of the 1928 trip. In it, he recorded meteorites sold and exchanged with various institutions. Figure 1.10 shows two pages from the notebook recording sales to the Buffalo Museum, Colgate University and Hamilton College. The notebook details other sales to Syracuse University, Hayden Planetarium, Clark University, Union College, the U.S. Museum, Cornell University and Case School.

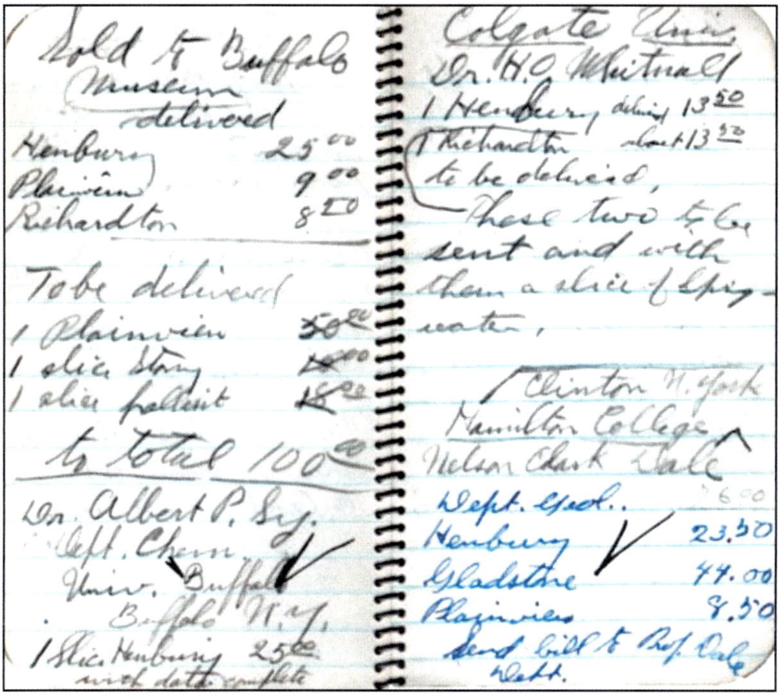
Sold to Buffalo
Museum
delivered
Henbury 25.00
Plainview 9.00
Richardton 8.50
To be delivered
1 Plainview 50.00
1 slice Stony
1 slice Pallasite
to total 100.00
Dr. Albert P. Sy,
Dept. Chem.
Univ. Buffalo
Buffalo N.Y.
1 Slice Henbury 25.00
with data complete

Colgate Univ.
Dr. H.O. Whitnall
1 Henbury, delivered 13.50
1 Richardton delivered 13.50
to be delivered,
These two to be
sent and with
them a slice of Spring-
water,
Clinton N. York
Hamilton College
Nelson Clark Dale
Dept. Geol.
Henbury 23.50
Gladstone 44.00
Plainview 8.50
Send bill to Prof. Dale

Figure 1.10. Pages of meteorite sales from Nininger's "Notes on trip" notebook. Courtesy ASU Buseck Center for Meteorite Studies.

While in Washington, D.C., he let students plan their own programs for the two-week stay in the nation's capital, except for certain assignments involving the U.S. National Museum and a few suggested points of interest. Nininger called on Dr. George Merrill, head curator from 1917 to 1929 of the Department of Geology, United States National Museum (now the National Museum of Natural History of the Smithsonian Institution). Nininger had consulted with him in 1923 concerning a Kansas meteorite that had been found. During this 1928 visit he presented Merrill with the idea of supporting a field program for the recovery of meteorites to add to the National Museum's collection. After a lengthy conversation, Nininger reports that Merrill responded: "Young man, if we gave you all the money your program required and you spent the rest of your life doing what you propose, you might find one meteorite." Nininger was disappointed but not deterred. He responded to Merrill: "the next time you see me I'll be selling you a meteorite." A few years later they met again and Nininger sold him two meteorites.

Figure 1.11. George P. Merrill. Courtesy Smithsonian Institution Archives, Record Unit 95, Box 17, Folder 7.

From Washington they traveled to Gettysburg, Philadelphia, New York City, and Boston. The year's work was now finished. Nininger noted that though the final days of the nine-month outing "lacked the gaiety of the start," later memories emphasized the brighter times and minimized the hardships. Looking back thirty-five years later, one of the students described it this way: "We were a motley crew. We were immature, but we learned to give and take. During the nine months all of us grew in certain personal virtues such as patience, tolerance, loyalty and cooperation." During thirty-six weeks, they had covered 19,000 miles, visiting the nation's capital, thirty colleges and universities, thirty-four museums, fifteen state capitol buildings, numerous libraries, a half dozen different kinds of mines, a score of industrial plants, and a dozen institutions of scientific research.

1930 Move to Denver

Determined to pursue this new "hobby," Nininger approached the President of McPherson upon his return and advised that he would be spending half his time in the search for meteorites. He thought it would have been unfair to retain the chair as head of the biology department and draw a full salary. The President responded: "You go ahead and spend the time on meteorites. We're glad and proud to have you do it. So give yourself a light load in teaching." This went on for six years until he decided that the generous arrangement would no longer be proper since his time on meteorites was increasing and his class time decreasing.

In February 1930, Nininger found that his interests in meteorite research was becoming difficult to resist. He wrote to the Colorado Museum of

Natural History in Denver. He wanted to loan his meteorite collection to the museum and suggested he might be able to work there. After discussing it with Addie, without waiting for a reply, they decided to move to Colorado even though he had no guarantee of a job. They had a 1929 Chevrolet and about $2,000 in cash. They loaded up their household goods and children into a borrowed farm truck. They moved into a row house near downtown Denver.

The McPherson newspaper ran an article about his move and it was picked up by the Associated Press. Dr. J. D. Figgins, Director of the Colorado Museum of Natural History, saw the article and contacted him with a job offer. He was offered a full-time position. He was also offered the option of a $50 monthly honorarium and use of the museum as his meteorite research headquarters. He would also retain ownership of his extensive collection if he exhibited it in the museum. Even though the pay was not good, he accepted the latter offer since it provided him with total freedom to pursue his passion of meteorite research and collecting. He became the museum's Curator of Meteorites.

Later that year, he opened *The Nininger Laboratory* in a spare room of a general store in Palmer Lake, Colorado, near the site of the Rocky Mountain Summer School. The location offered an empty room, and more importantly, it held the saw that he had built during the summer school. With the saw he was able to set up a meteorite cutting laboratory. This was the first commercial cutting laboratory for meteorites in the country. The only other equipment at the time was at the National Museum in Washington, D.C. He changed the name of the lab in 1937 to *The American Meteorite Laboratory*.

Figure 1.12. *Life* magazine, September 6, 1937, photo with caption: "Cutting a slab from a hard, stony meteorite. Onto the slow-moving toothless steel band trickles a stream of water carrying a film of carborundum powder. This very powerful abrasive is used by Dr. Nininger to do the actual cutting." ((c) Life Picture Collection/Shutterstock.

1933 Society for Research on Meteorites

Frederick Leonard (1896–1960) was an American astronomer and faculty member at the University of California Los Angeles (UCLA). He founded the Department of Astronomy in 1931 which he headed until his death in 1960.[6] Nininger first became acquainted with him through correspondence in 1930 concerning the purchase of meteorites.

Figure 1.14. *Life* magazine, September 6, 1937, photo with caption: "Polishing a meteorite on a buffer in the American meteorite Laboratory. Meteoriticists study structural composition on smooth surface." (c) Life Picture Collection/Shutterstock.

Leonard advocated to establish an organization to study meteoritics. When the two met in 1932, Leonard pursued the idea with him. After some hesitation, Nininger eventually agreed to assist in this effort. An organizational meeting was held in August 1933 at the Chicago Field Museum. Leonard was elected president and Nininger became the organization's first secretary-treasurer. Four years later, Nininger was elected president.

Within five years, the Society had doubled in size, with members from the United States and ten other countries. Nininger retired as president in 1941. Meetings were suspended during World War II (1942 through 1945). When reconvened in 1946, the organization's name was changed to The Meteoritical Society. Today the Society has a membership of over 1,000 from 52 countries with interests in meteorites, cosmic dust, asteroids and comets, planets, impacts, and the origin of the Solar System.

Meteorite Leaflet

Over the next several years, Nininger continued to "educate" the public about meteorites by describing what they looked like, often with samples, and offered cash if they found any. He produced a leaflet for this purpose (Figure 1.15). For example, Nininger's notebook on the search for the Akron and Holyoke meteorites (Figure 1.16) shows the distribution of 800 leaflets in one week, and 600 the following week, while visiting 20 towns.

Note the time of the occurrence and listen for any explosion-like, or thunderous, sounds, noting the time they were heard.

Draw two lines on a paper, one representing the horizon, the other the line of flight of the meteor, thus showing the direction of flight and the angle of descent.

Estimate in degrees the approximate height above the horizon at which the light vanished. It is 90° from the horizon to directly overhead. A measurement of approximately 15° can be obtained by extending an arm full length and using the width of the palm plus the extended thumb as a measure.

Write immediately to the American Meteorite Laboratory, P. O. Box 2098, Denver, Colorado 80201, giving name and address, an estimate of the size of the meteor as compared with another object in the sky such as Venus or the full moon, and the information listed above.

When a sufficient number of witnesses report the sighting of a fireball, instrument measurements can often be made by a trained meteoriticist and meteorite specimens recovered through an intensive field survey.

A most important phase of today's research is the study of meteorites less than one month old—or, ideally, less than one week old—on earth.

Keep this leaflet handy. You may someday have the opportunity to make a vital contribution to America's scientific program.

AMERICAN METEORITE LABORATORY

P. O. BOX 2098 — DENVER, COLO., 80201

Meteorites are Valuable For Research

How to Recognize Them

A survey during recent years has demonstrated that on an average less than one person in a thousand is able to recognize meteorites in their natural condition.

Meteorites are not light porous rocks. They are often marked with shallow pits but are not porous.

They are not round like a ball. At least no round one is yet known.

They are not hollow.

Meteorites do not come to the ground in a burning condition. They do not set fires. They burn while high in the air but generally cease burning about 5 to 20 miles above the soil.

Meteorites are usually much heavier than ordinary rocks.

They are generally irregular in form—almost any shape. They are often pitted more or less. The corners and edges are notably dulled or rounded. A few meteorites are conical in shape.

Meteorites are covered with a thin fusion crust due to burning during their flight through the atmosphere. This crust is nearly or quite black in most meteorites at the time of fall. Later it becomes brown from rust if exposed to the weather.

They nearly always contain an alloy of nickel-iron. This metal may be in small grains embedded in a stony matrix or it may constitute almost the entire meteorite. In either case it can be detected by grinding a corner of the suspected specimen with an emery wheel or emery paper which will reveal chrome-bright metal.

Meteorites are important for research purposes and good prices are paid for them. Any specimen which conforms to the above description and meets the emery wheel test should be submitted for other more exhaustive tests at the hands of a specialist. The American Meteorite Laboratory makes such tests free of charge; however, it is requested that postage be sent for return of the specimen should the sender wish it back regardless of whether it is or is not meteoritic.

Detach a sample not smaller than a walnut from a large specimen (small specimens should be sent intact) at some place where such detachment will do the least damage and send it for examination to the American Meteorite Laboratory, P. O. Box 2098, Denver, Colorado 80201.

Should tests prove a specimen to be meteoritic, an offer of purchase will be made.

Meteorites thus acquired become objects of study in scientific laboratories throughout the world.

If You See a Fireball

The light which a meteorite makes as it passes through the atmosphere is called a **meteor.** When a meteor appears somewhat larger than the planet Venus or approaches the size of the full moon, it is called a **fireball.**

If you should witness the passage and breakup of a fireball, note the exact direction in which it disappeared and select a landmark as far distant as possible which is in direct line with its vanishing point. Note carefully your own exact location.

Figure 1.15. This leaflet, created by son-in-law Glenn Huss, is similar to the leafelet distributed by Nininger to promote the education and collection of meteorites among the public. Exterior portion on left and interior on the right.

1933 The Mines Magazine

In addition to submitting many articles to *Popular Astronomy,* Nininger would submit articles to a variety of magazines in hopes that his story would be picked up to further the discovery of meteorites. The first published article written by Nininger, that could be found, appeared in the December 1933[7] edition of *The Mines Magazine*, published by the Colorado School of Mines since 1910. The article was entitled "Meteor Craters vs Steam Blowouts." He stated that the "purpose of this paper is to point out evidence in support of the meteoric hypothesis as against the blowout hypothesis as an explanation of the origin of a certain type of depression or crater that is not uncommon. . . ." He discusses the crater at "Coon Butte, Arizona" and at Odessa, Texas (both discussed in Chapter 2), among others. His arguments concerning meteoric craters are covered later in Chapter 3 on Meteor Crater, but in this article we see the beginning of his thoughts on the subject.

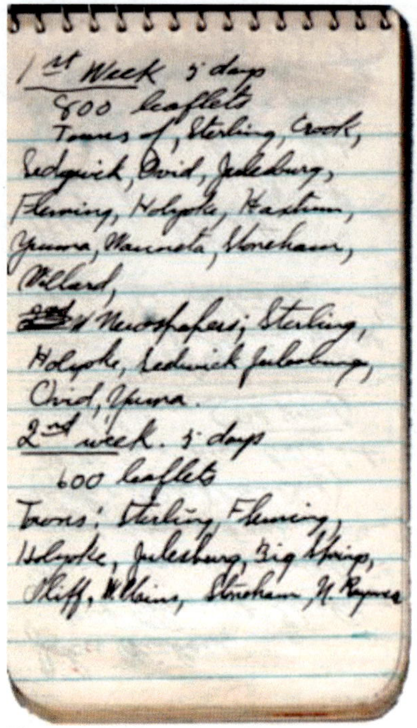

Figure 1.16. Page from Nininger notebook of 1940 visits to Colorado towns. Courtesy ASU Buseck Center for Meteorite Studies.

A second article by Nininger appeared in the January 1935[8] issue. This one-page article was entitled "How to Recognize Meteorites."

1934 Literary Digest Magazine[9]

Nininger's next article appeared in the March 17, 1934, issue of *The Literary Digest*, a general interest weekly magazine, with a circulation of over one million. Written by Nininger, the article was entitled

Figure 1.17. *The Mines Magazine* cover of December 1933. Reproduced with permission.

"When the Sky Rains Stone and Iron." It discussed various meteor showers, and in particular the Leonid shower of 1833 and its subsequent visits. He described the cause of meteor showers and the creation of craters when meteors reach the earth. Continuing his practice of teaching the public how to recognize meteorites, the article described their color, weight and how it may "be pitted like a lump of putty into which somebody has pushed his thumbs and fingers."

1937 Life Magazine[10]

Life was an magazine published weekly from 1883 to 1972, and as a monthly from 1978 until 2000. During its golden age from 1936 to 1972, *Life* was known for the quality of its photography.

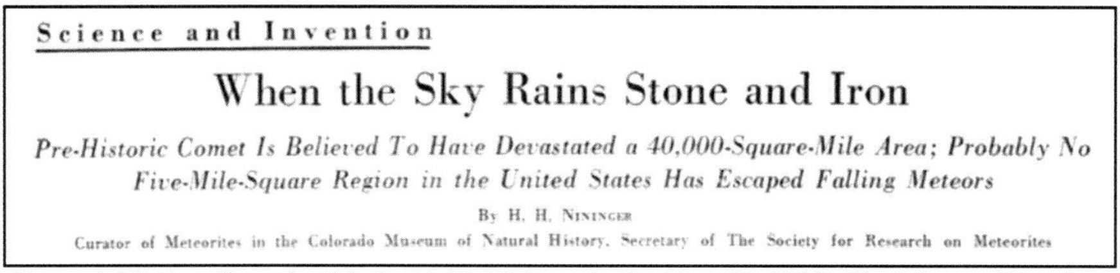

Science and Invention

When the Sky Rains Stone and Iron

Pre-Historic Comet Is Believed To Have Devastated a 40,000-Square-Mile Area; Probably No Five-Mile-Square Region in the United States Has Escaped Falling Meteors

By H. H. Nininger

Curator of Meteorites in the Colorado Museum of Natural History, Secretary of The Society for Research on Meteorites

Figure 1.18. Headline of article by H. H. Nininger in the May 17, 1934, *The Literary Digest*.

An article about Harvey and Addie appeared in the September 6, 1937, issue of *Life*. The article was titled "Chasing Meteorites is a Career." It covered three pages with 16 photographs of Harvey and Addie, meteorites and his discoveries. It stated that "whenever a meteorite does strike the Earth, the chances are that the world's most persistent star-dust chaser, Dr. Harvey Nininger of Denver, will discover it, buy it and sell it to a museum." Images in the article include those (Figure 1.19 and 1.20) showing Nininger on the road talking about meteorites.

As with many of these articles, the intended result was to get leads on a meteorite. Nininger retained many letters from individuals responding to the articles. In this case, he received a letter from a Mr. Frank Post of Palo Verde, Arizona, in which he says "I read your article in Life magazine. . . while prospecting a number of years ago I ran across a heavy black substance which I thought to be a meteorite." Many times Nininger would write "Investigate" on the top but this particular letter contains no such notation so it is unclear if he pursued this lead or not.

1941 Roving Reporter[11]

Ernie Pyle was a Pulitzer Prize–winning journalist and war correspondent who was best known for his stories about ordinary American soldiers

Figure 1.19. This image from *Life* shows Nininger (white hat) outside their trailer. The caption reads: Here he is describing meteorites to farmers, asking them to send him their finds. (c) Life Picture Collection/Shutterstock.

Figure 1.20. This image from *Life* article inside the trailer. The caption reads: Harvey Harlow Nininger passes all his time hunting meteorites and lecturing on these rare witnesses from outer space. (c) Life Picture Collection/Shutterstock.

during World War II. He also wrote as a roving human-interest reporter for a newspaper syndicate under the heading "The Roving Reporter." This earned him wide acclaim for his simple accounts of ordinary people.

FRIDAY, JULY 18, 1941

THE ROVING REPORTER

By ERNIE PYLE

Dr. Nininger Will Quit Chasing Meteors When One Hits His Head

Figure 1.21. Column header for *The Roving Reporter.*

On Friday, July 18, 1941, his column featured an article entitled "Dr. Nininger Will Quit Chasing Meteors When One Hits His Head." This appeared in the *Rocky Mountain News* and more than 50 other newspapers. Pyle wrote that Nininger "probably never had such a pumping in his life as he got from me. But he had all the answers." He asked about people getting hit by meteorites, the nature of a meteorite and whether they were all iron. He described how Nininger was probably away from home "about half the time" searching for meteorites, although the whole family occasionally went on the chase when a fall was reported. He concluded the column with: "He doubtless will continue to be America's No. 1 meteorite-sleuth until one hits him on the head."

1941 Ford-TWA Meteorite Survey Expedition[12]

TWA Stratoliner pilot, Captain H. H. Holloway was traveling between Los Angeles and Albuquerque. One of his passengers was Edsel Ford, son of Henry Ford of Ford Motors. Looking down, Holloway had spotted a series of deep gashes in the earth. Halloway suggested to his passenger that the craters might be meteorite pits. Since just 45 miles away was Meteor Crater, he thought these craters may have been produced by "chips" from the meteor that created the larger crater. If so, Holloway suggested that deposits of meteoritic nickel-iron might be found in these cavities.

A joint Ford-TWA expedition was organized. Nininger was selected to be the expedition leader. After visiting the craters and spending many hours examining them, Nininger concluded that rather than being formed by meteorites, they were rock-roofed caverns that collapsed when the meteorite that created Meteor Crater struck the earth. He explained that when the meteorite smashed into the earth, "a terrible earthquake was set in motion. This collapsed the cavern roofs in their weakest points, resulting in these jagged pits." It was established that Meteor Crater, and

Figure 1.22. Members of the Ford-TWA Meteorite Survey Expedition arrive at Winslow, Arizona, by regular TWA transcontinental planes. *Ford News* photo.

Figure 1.23. *Ford News* photo with caption: "Winding across the rough northern Arizona desert plateau, the caravan of Ford passenger cars and station wagons receive a real test."

Figure 1.24. 1941 Ford-TWA Meteorite Expedition on the survey. *Ford News* photo.

these pits, were created at the same time based on the similarity in the degree of weathering of the limestone formations.

Figure 1.25. *Ford News* photo with caption: "An aerial view of some of the thirty-two newly discovered craters on the Arizona plateau emphasizes their resemblance to great shell holes. In the background is the Chevalon Canyon where entrances to the hidden subterranean caverns may be found."

Strange As It Seems[13]

Strange as It Seems appeared as a syndicated cartoon feature published from 1928 to 1970, and became a familiar brand to millions around the globe for its comic strips, books, radio shows and film shorts. Created by John Hix, *Strange as It Seems* was distinguished for its adherence to Hix's standard that every published fact be verified by a minimum of three sources. In Hix's words, *Strange as It Seems* is a library of "the curious, in nature and humankind, set adrift on the vast sea of public opinion with the hope that it will fulfill its mission to entertain and acquaint its viewers with some of the marvels of the world in which we live." The cartoon

Figure 1.26. *Strange As It Seems* featuring H.H. Nininger. Date unknown. Courtesy Sedona Historical Society.

shown in Figure 1.26 appeared in a number of publications but the exact date is unknown. The cartoon was contained in material provided by Nininger to the Sedona Historical Society.

Ripley's Believe It Or Not[14]

Cartoonist LeRoy Robert Ripley published his first "Believe It or Not!" cartoon in the *New York Globe* on December 19, 1918. With a positive response from readers, the cartoon began appearing weekly. Ripley claimed to be able to "prove every statement he made" because he worked with a professional fact researcher who assembled an array of odd facts and who also verified the claims submitted by readers. Subjects covered in Ripley's cartoons and text ranged from sports feats to little-known facts about unusual and exotic sites.

Figure 1.27. *Ripley's Believe It Or Not* featuring H.H. Nininger. Date unknown. Courtesy Sedona Historical Society.

The cartoon shown in Figure 1.27 appeared in a number of publications but the exact date is unknown, although the particular newspaper in which this cartoon appeared stopped publication in 1950. The cartoon was contained in material provided by Nininger to the Sedona Historical Society.

1942 Desert Magazine[15]

Desert Magazine was a monthly publication that focused on the desert country of the Southwestern United States and Northwestern Mexico. It covered a broad range of desert subjects including regional travel and exploration, the visual arts, cultural history, natural history and other topics. It was published between 1937 and 1985.

Nininger appeared in four issues, starting with an article entitled "Visitor From a Distant Planet" in the July 1942 issue. Authored by Nininger, the article is a colorful presentation on meteorites. In describing a descending meteorite:

> *Sparkling and sputtering in a racing fountain of flaming incandescence, it thunders downward through the resisting sea of air*

with the noise of a thousand dive bombers.

He describes Meteor Crater as "The World's Greatest Bomb Crater."

> *Imagine the Rose Bowl or the Yale Bowl enlarged to seat 2,600,000 people and you can visualize the extent of the great Barringer meteorite crater in Arizona.*

Most of the article discusses craters formed by meteorites with equally colorful descriptions:

> *These super-charged particles now leap with twenty times the violence of exploding T.N.T., shattering and lifting the surrounding rock strata, and hurling skyward millions of tons of fragments. These fall in a crushing shower . . . .*

This article perhaps gives a hint of why publishers of many magazines felt that Nininger's accounts would find an interested public with his compelling and colorful descriptions.

1942 Esquire Magazine[16]

Esquire magazine was founded in 1933. It flourished during the Great Depression and World War II. It was supposed to have a quarterly press run of a hundred thousand copies, however, demand was so high that by its second issue (January 1934), it transformed itself into a more refined periodical with an emphasis on men's fashion and contributions by Ernest Hemingway, F. Scott Fitzgerald, and others. The May 1942 issue featured an article entitled "Miner of Meteorites" by Roscoe Fleming. Fleming began his newspaper career in Fargo, S.D., and worked for many papers before moving to Denver in 1936 to work for the *Rocky Mountain News*. He became a columnist for The *Denver Post* in 1941. This is probably how he became acquainted with Nininger at the Denver museum.

"We bump for an hour through the snow-powdered winter woods of the Sangro De Cristo foothills, over a villainos, semi-abandoned mine road that winds along the mountains through an endless series of barbed-wire ranch gates and leads us through fords where the water laps the running board." He is travelling to meet Nininger, whom he describes as "the world's foremost meteorite sleuth and miner, making his living a strange, hard way."

Nininger is investigating a possible meteorite crater that came to be known as the Crestone (Colorado) Crater (Figure 1.28). The Crestone

Figure 1.28. Crestone Crater. Courtesy of the American Meteorite Laboratory Photo Archives, Collections Research for Museums, Denver, CO.

Crater is an elliptical bowl measuring 355 ft by 246 ft with a mean depth of 23 ft. Some fifty years ago, a fireball was observed over the area but a meteorite was never found. Thinking there may be a meteorite crater, Nininger flew over the area in 1941 looking for a possible location.

The article provides background on Nininger and his quests for meteorites. For this expedition, Nininger took aerial photographs, did ground searches for meteorite fragments, and dug shafts in the crater floor. No meteorite fragments were found, so Nininger suggested it may have been formed by an icy comet fragment. Even Daniel Barringer of Meteor Crater (who will be discussed in Chapter 3) described it as a suspected meteorite crater. Other studies of the crater and surrounding area, likewise, failed to find any meteorite fragments, but this article added to his mystique as a meteorite hunter.

1942 Arizona Highways[17]

Arizona Highways is a magazine that contains travelogues and artistic photographs related to the State of Arizona. It is published monthly by a unit of the Arizona Department of Transportation. The May 1942 issue featured an article entitled "Arizona's Newest Natural Wonder." It covered the 1941 Ford-TWA Expedition mentioned earlier. The article begins with a grandiose claim:

> *After 25,000 years of unseen existence, a newly discovered topographical feature near the edge of the northeastern plateau has been explored, and indications are strong that the Sunshine State may*

boast underground caverns of a size with Kentucky's Mammoth Cave, and of the grandeur of Carlsbad Caverns of New Mexico.

The article describes the expedition of "20 men" making a laborious trip to the craters. Upon arrival, they explored the caverns on horseback. It goes on to quote Nininger as suggesting that between these craters are gigantic subterranean openings of "perhaps unrivalled size."

Figure 1.29. Dr. Nininger in the cockpit of one of the planes that surveyed the plateau. Courtesy *Arizona Highways Magazine.*

The article describes Nininger's belief that the craters were sinkholes that opened upon the earthquake effects of the impact of the meteorite that created Meteor Crater. Subterranean caverns, however, were never found.

1943 Collier's Magazine[18]

Collier's was a general interest magazine founded in 1888 and ceased publication in January 1957. *Collier*'s had a circulation of over two million, but it began to lose readers during the post-World War II years. The December 11, 1943, issue carried an article entitled "Meteorite!" by Harvey Nininger and Roscoe Fleming. The article is a primer of sorts on meteorites - what they are, how they travel through the universe and what happens when they reach earth. There is a short discussion of "the great Barringer meteorite pit." They mention that meteorites may contain diamonds, gold, platinum and how "life itself may have been brought to the earth in the form of spores traveling either by themselves or embedded in meteoritic fragments."

1947 Desert Magazine[19]

Desert Magazine included an extensive article in the May 1947 issue entitled "Star Dust Hunter."

The article was an excellent biographical overview and presents Nininger's new American Meteorite Museum near Meteor Crater (Figure 1.30 and discussed in Chapter 4). The article describes how visitors receive a short lecture on how to tell the difference between a meteorite and terrestrial rocks, and how to make "intelligent" reports of meteorite falls.

Figure 1.30 Photo of Harvey and Addie Nininger outside the American Meteorite Museum in the May 1947 *Desert Magazine* article.

Figure 1.31 Photo of Harvey Nininger in the May 1947 *Desert Magazine* article.

1947 Liberty Magazine[20]

Liberty Magazine was founded in 1924 and, at its height, had an ongoing circulation of more than 3 million a week. It was founded on the high quality and originality of its art, stories and features. It was thought to be the second most popular magazine after the *Saturday Evening Post*. It ceased publication in 1950.

The January 11, 1947, issue[21] included a feature article written by Roscoe Fleming. Fleming, discussed in earlier articles, was a journalist but also an author who began publishing works of science fiction interest. Some of his science fiction bent shows in the title of the article - "Bombs from the Blue." The focus of the article was on the October 9, 1946, meteor shower from Comet Giacobini-Zinner. In spite of the presence of a full moon and city lights, the shower was observed by thousands of people in the western United States. It described the passing meteor as "a large blue-white ball with white light back of it. It left a yellow train which lasted over three minutes, spreading out and drifting into the shape of a horseshoe . . . with the maximum frequency [of meteors] occurring between 7:30 and 8:00 p.m., probably exceeded 10,000 meteors per hour."

The article described meteor showers and why we do not read "of houses pierced, skyscrapers knocked down, and people and animals killed by such terrible missiles." He goes on to describe the meteor that created Meteor Crater as "the great American meteorite" and the most sensational "known" example of a great meteorite. It continues:

> "We say 'known' because Dr. H. H. Nininger, our foremost meteorist (who has recently established a museum at the Winslow crater with his collection of meteorites, the most extensive in the United States), believes that the oceans, vegetation, and erosion on land have masked from us the more ancient scars of even greater meteorites."

As often happened, the article resulted in letters to Nininger such as a letter[22] dated March 9, 1949, from Tucson that read: "I read your name and address in *Liberty* magazine January 11, in an article on meteorites, and I think I know where one is or two maybe." Nininger catalogued the letter but there is no notation as to whether he followed up - but the article had its intended effect of educating the public and reporting possible finds. [Note: Comet Giacobini-Zinner is a periodic meteor shower that returned in 2018 with less spectacular results.]

1947 Society of Vertibrate Paleontology Fieldtrip[23]

The 1947 Society of Vertebrate Paleontology summer excursion focused on the continental Triassic and Early Jurassic deposits in northern Arizona. In 1938, Samuel Welles discovered the Meteor Crater Quarry (near Meteor Crater) which has the richest accumulation of vertebrates in the area. The discovery of the quarry sparked more investigations of the area that had previously been regarded as barren of fossils.

As mentioned earlier, Nininger's original area of study was biology. He had also undertaken several field trips with students in search of fossils. On the Society's field trip, members of the group visited Meteor Crater for a lecture by Dr. Nininger (Figure 1.32).

1948 Saturday Evening Post[24]

Figure 1.32. Nininger addresses members of the Society on the rim of Meteor Crater. Courtesy of Northern Arizona University Special Collections Archives.

The *Saturday Evening Post* was a weekly publication that began in 1897. In its heyday, it was one of the most widely circulated and influential magazines, reaching about two million homes every week. The magazine was known for its cover art by Norman Rockwell who painted 322 covers between 1916 and 1961. Today it is owned by a nonprofit organization that preserves its legacy.

The April 17, 1948, issue contained an article entitled "Star Chaser." The article is a short biography of Nininger that traces the origin of his interest in meteorites and how he found meteorites. It describes, in particular, his searches for the Paragould and Hugoton meteorites (Chapter 2). The author summarizes his successes by observing "he has no near competitor in that odd field." One colorful passage observes:

> "Nininger is a scientist, but not out of a copybook. More orthodox operators sometimes look down their noses at him, and wonder how the devil he gets so many meteorites. The fact is, besides knowing what's in the books, he has a lot of what Sherlock Holmes had. He follows clues and scents with uncanny skill."

1948 Desert Magazine[25]

The second article in *Desert Magazine* in which Nininger appeared was in the March 1948 issue. It was written by Nininger and was entitled "On the trail of the meteorites." The article details a witnessed meteorite fall on October 31, 1947, on the nearby Navajo Reservation. A couple who had recently visited the museum rushed in to report that they had

Figure 1.33 *Post* photo caption: "In addition to finding meteorites, at which he is a champion, Nininger buys a good many." Photos © SEPS licensed by Curtis Licensing Indianapolis, IN. All rights reserved.

Figure 1.34. *Post* photo caption: "Near the awesome Meteor Crater, in Arizona, Nininger hunts for buried meteorites with a war-surplus mine detector. His wife follows with shovel, pick and magnet." Photos © SEPS licensed by Curtis Licensing Indianapolis, IN. All rights reserved.

witnessed a fall about 5:30 pm the previous evening. While relating their account, a second couple came in to report that they saw it as well. Nininger reported that "All day, museum visitors reported having seen the phenomenon."

The next day, Harvey and Addie drove to Gallup, New Mexico, where they met two Navajo witnesses. They drove north from Gallup where they met a Navajo guide who had also seen the fireball. Within 48 hours, based on these accounts, Nininger plotted the probable location of the fall (Figure 1.35). Nininger admitted that "these surveys are never exact. Allowances of three or four miles variation from the indicated course must be made."

Nininger admitted that they did not find any meteorites during their visit, but he provided advice to any collectors who might wish to try their luck. As of this writing, no fragments from this fall have been reported.

1948 Arizona Highways Magazine[26]

An article in the May 1948, issue of *Arizona Highways* was entitled "On the trail of the meteorites" written by Nat McKelvey.

Based on the title and content, it appears that Nininger sent his article to both *Desert Magazine* and *Arizona Highways*. *Arizona Highways* assigned the story to Nat McKelvey and photographer Tad Nichols. It is interesting that the article retained Nininger's title but it was turned into

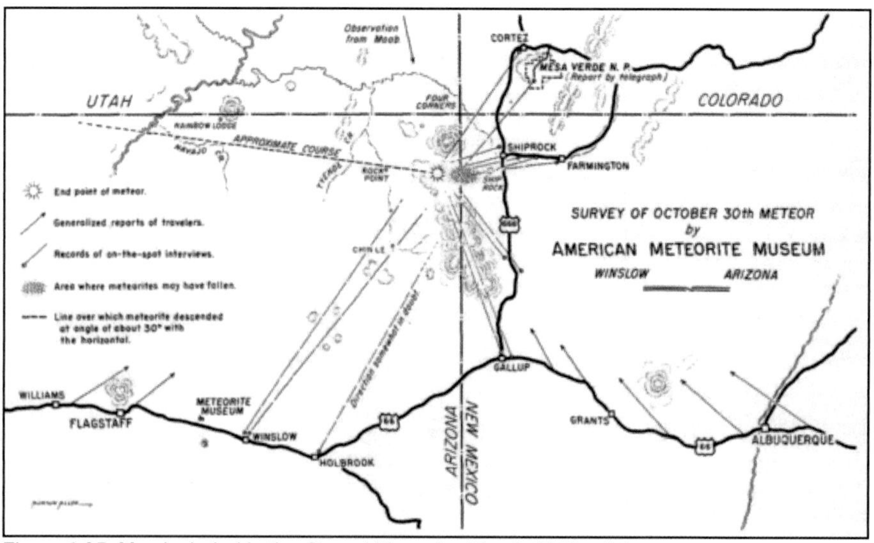

Figure 1.35. Map included in the *Desert Magazine* March 1948 article.

a biographical piece with no mention of the meteorite fall in the original article. There are several pictures of the museum (Chapter 4) which probably led to more visitors.

Figure 1.36. Image from 1948 *Arizona Highways* Magazine.

1951 Newsweek[27]

Nininger spent many years studying Meteor Crater (Chapter 3). The May 7, 1951, issue of *Newsweek* included an article entitled "Meteorite Mystery" about Nininger and Meteor Crater. The article described how Daniel Barringer spent nearly $1,000,000 from 1904 to 1931 drilling for the iron mass that he thought created the crater. But "in 1938, a sedate, sandy-haired man from the Colorado Museum of Natural History in Denver shocked searchers by asserting that they were wasting their time." Nininger theorized that the meteorite "had literally gone up in thin air." The article went on: "This week meteoriticist Dr. Harvey Harlow Nininger had in hand the first proof of his theory."

Nininger spent the previous three years going through soil samples collected from 135 locations in and around the crater with his magnetic "rakes." The article quoted him explaining that "There cannot be a meteorite mass here . . . because the meteorite exploded on the instant of impact. It became a huge ball of gases (metal vaporized by heat), which condensed in the air over the crater area. Then the spheroids were formed, and they rained to the ground."

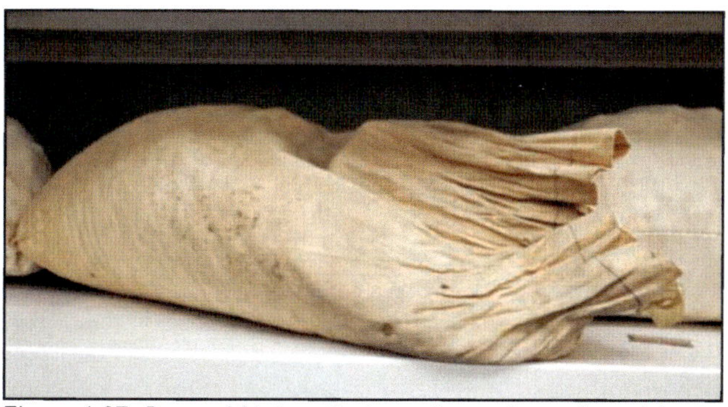
Figure 1.37. Bags of Meteor Crater soil samples collected by Dr. Nininger now in the vault of the ASU Buseck Center for Meteorite Studies. Used with permission.

The article explained that Nininger planned to study the soil sampling (Figure 1.37) for another year, after which he hoped to prepare a new estimate of the original meteorite's size, "the first based on material remains."

1952 Argosy Magazine[28]

Argosy was launched in 1888 and was published in various formats and genres until 1978. The June 1952 issue contained a feature article entitled "The Man Who Chases Meteors" by Morton M. Hunt.[29] Hunt was a well-known freelance writer from 1949, specializing in the social and behavioral sciences. The sub-heading of the article was "Flying fragments from outer space changed Dr. Harvey Nininger's life."

Figure 1.38. *Argosy* Caption: "Nininger uses electronic metal detector to locate meteorites."

This is another short biographical piece, very similar to the *Post* and other publications. With many similar stories, it would appear that these authors picked up on an article submitted by Nininger and used it as the basis of their piece. He writes that Nininger "is probably the only man who gets his living from outer space." This was one of the first articles that mentions the use of electronic devices (Figure 1.38), noting that "today he swings a regular GI mine detector when he goes prospecting."

1961 Point West Magazine[30]

Point West magazine of Arizona was a short-lived upscale magazine for Arizona including articles on fashion, art, dining, entertainment, etc. It boasted that it was "edited for people who care about the past and future of our state." The June 1961 issue carried a five-page article entitled "The greatest natural laboratory in the world" featuring a full-page image of Meteor Crater. The article's subtitle was "Harvey H. Nininger, Researcher of the skies, collector of natural treasure."

The focus of the article was the acquisition of the Nininger Collection by Arizona State University, with some biographical information. In describing Nininger:

> "Preoccupied with deducing from them the mysteries of the universe, this wiry, witty scientist of seventy-odd has for more than three decades been obsessed by collecting and examining these weighty messengers from without. He has maintained his curiosity through physical, professional and financial hardship."

During the interview, in his "courteous, country-boy, straight-forward midwestern, matter of factness" Nininger lamented that he wished that he had more time for pure research. That is why, he said, he especially enjoyed his work "at the great Meteorite Crater outside Winslow, where he spent a long period of research." The article closes by noting that the collection will be a lasting tribute to Nininger, "a man with a brilliant mind, backed up by a strength of purpose."

1962 Saga magazine[31]

Saga magazine was a men's interest magazine focusing on Western adventure stories. The December 1962 edition featured a five-page article on Nininger entitled "Millionaire Miner of the Milky Way" by Theodore Berland. Theodore Berland (1929-) is author of more than 20 books and hundreds of articles in magazines and newspapers. The article begins by relating an incident in October 1957 when a Navy flier over Myton, Utah, spotted an object flashing across the sky that was so close that he had to veer his plane sharply to avoid it.

This was the same month the Space Age began with the launch of Sputnik. Nininger had received a call from the editor of a Colorado newspaper who wanted to know "what was it? Sputnik, a missile or what?" Nininger told him that it sounded like a meteorite. This is another very good biography tracing his career from McPherson, to Meteor Crater, to the sale of his collection to Arizona State University. Summarizing his career, the article ends with a quote from Nininger:

> "I still don't see how I ever did it. I don't see how a young fellow with a family could ever have dared to take that chance. But I guess I was more daring than even I realized. And I made a pretty good living, after all."

1975 Denver Post Empire Magazine[32]

Empire Magazine was a Sunday supplement of the *Denver Post*. The cover (Figure 1.40) of May 25, 1975, highlighted a lengthy article entitled "How Harvey Nininger's world was created." Nininger was interviewed for this article at his Sedona home (Figure 1.39). This biographical piece, by staff writer Zeke Scher, covered highlights of his career and pointed out that "Nininger showed the gusto of his pioneer parents as he dug into the new science with few guidelines and few sources of knowledge. And with only his personal assets to fund the program it was a hand-to-mouth operation." He is credited with making the "science of meteoritics a household word."

The article relates that Nininger worked at the Colorado Museum of Natural History (now the Denver Museum of Nature and Science) starting in 1930 and for the next 15 years. During this time he was able to pursue his "impossible dream" of hunting meteorites. During his time at the Museum, he accumulated more than 1,500 specimens from some 200 different falls. He added 31 new meteorites to the official list in 1937 alone.

The article quotes Dr. Fred Whipple, Harvard professor of astronomy and director emeritus of the Smithsonian Astrophysics Observatory, who said that "Nininger and meteorites go together like bread and butter."

Figure 1.39. Dr. Nininger on the porch of his Sedona home. Courtesy of Sedona Historical Society.

Father of American Meteoritics

Because of his pioneering work, some of which has been described in these publications, Harvey H. Nininger is considered by many today to

Figure 1.40. May 25, 1975 cover of *Empire Magazine* of the *Denver Post*.Courtesy of Sedona Historical Society.

be the "father" of American meteoritics. He followed his passion and beliefs and persevered despite the skepticism in the scientific community at the time. Most people would probably have changed direction if they were confronted by Dr. Merrill at the United States National Museum who tried to discourage him in the search for meteorites. Harvey Nininger was a pioneer of modern meteorite research. He led the way when others, like Dr. Merrill, thought the effort was frivolous. Nininger's collecting methods, cataloging, and displaying of meteorites are in many ways still the standard.

Even after his death in 1986 at age 99, Nininger was still showing up in the literature. The September 1986 issue of ***National Geographic*** included an article[33] entitled "Meteorites: Invaders From Space." In discussing the difficulty of finding meteorite falls, the author credits Nininger for having developed "the most successful method of tracking such falls."

The September 1989 issue of ***Smithsonian*** magazine[34] included an article entitled "Stop to consider the stones that fall from the sky." The article discusses the history of meteorite falls and how they were regarded by the scientific community over the years. The idea that rocks would fall from the sky was ridiculed for many years. When it was eventually agreed that meteorites existed and came from space, there was still great skepticism about meteorites creating craters on Earth. Even Arizona's famous Meteor Crater was initially thought to have been caused by a steam explosion from within the Earth. The author points out that "throughout the first half of the [20th] century, geologists expended extraordinary efforts trying to explain away evidence that craters on Earth had been formed by impacts of extraterrestrial bodies." In one instance, when a geologist came upon a crater with iron stones, he announced that "prehistoric people had dug the pits to bury the iron."

The *Smithsonian* article states that "The uncompromising rejection of the presence of craters on Earth began to change with the work of Harvey Harlow Nininger." As discussed in the next two chapters, Nininger was convinced of the presence of craters and sought them out. His first discovery of a crater was near Brenham, Kansas, at what the farmer called a "buffalo wallow." Yet the article points out that in the 1950s, students were being taught that the only meteor crater on Earth was Meteor Crater in Arizona. "That tunnel-vision began to change with Harvey H. Nininger."

CHAPTER 2

The Star Chaser

Gentlemen, I would rather believe that those two Yankee professors would lie than believe that stones would fall from heaven.
—- Thomas Jefferson[1]

The Beginning in 1923

As mentioned in the previous chapter, Nininger was standing outside with a fellow professor at McPherson College on November 9, 1923. "Suddenly a blazing stream of fire pierced the sky, lighting the landscape as though Nature had pressed a giant electric switch," Nininger recalled. "The blade of light vanished . . . leaving a darkness seeming thicker than before."

Using the technique that would serve him well the rest of his investigative life, Nininger wrote to the local papers and asked for information from others who had witnessed the meteor. He sifted through the many reports that came in for the most helpful information, and combining it with his own observation, estimated that it had landed somewhere in the vicinity of Coldwater and Greensburg, Kansas. He began to search the area on weekends with small volunteer groups of students.

With each excursion he would talk to the local press as well as community and church groups. He would give lectures in rural high schools and grade schools, hoping the farm children would spread the word among their parents.

During one of his weekend searches he was approached by two farmers who had several small meteorites from a recognized fall twenty years earlier. He managed to purchase them from his meager family budget. These burned a deep interest to learn more about meteorites.

Nininger went to the geology department at the University of Kansas at Lawrence with the specimens, but they knew nothing of meteorites. He then went to the astronomy department. He talked to a professor who was, likewise, of little help. The subject of meteorites was simply not being taught. He described his mood after his visit to the university as both appalled and challenged.

He wrote to the Field Museum of Natural History in Chicago for some information since they had a sizeable collection of meteorites at the time. In return, he was sent a copy of O.C. Farrington's *Catalogue of the Meteorites of North America* published in 1915. He learned that the first recorded meteorite find was only made in 1803. Farrington's catalogue noted: "The total number of meteorites recognized by the writer in North America up to January 1, 1909, is 247."[2] Nininger assumed that the reason for the lack of interest was due to the scarcity of meteorite finds. He was convinced that he could improve those statistics if he looked in enough places and talked to enough people. As we will see later, the collection, described in his 1950 catalog, included meteorites from 587 falls. Below are some of his more notable or interesting finds.

1924 Coldwater Meteorites

Over the next year Nininger was able to recover two previously unknown meteorites. During one of his excursions to the area, he stopped by and visited a local church during services. He was allowed to address the congregation about his hunt. After the service, the Deacon approached him to suggest that he may have a meteorite in his backyard. He showed Nininger a 41 lb (18.6 kg) iron meteorite that he had plowed up in his field. Nininger purchased the meteorite from the Deacon on the spot for one dollar a pound. This was referred to as the **Coldwater (iron)** meteorite. In recounting its discovery, Nininger said that "the thrill that it gave me can never be described."

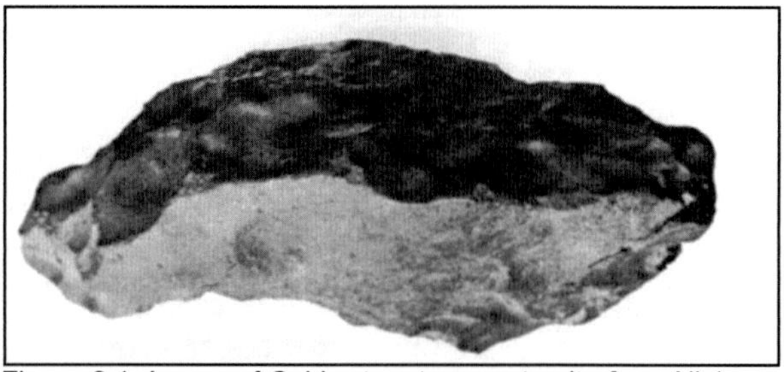

Figure 2.1. Image of Coldwater-stone meteorite from Nininger Collection in *Meteorites: A Photographic Study of Surface Features Part 1. Shapes*, p. 18.

The second discovery was an 11 lb (4.99 kg) meteoritic stone that a farmer had discovered during his plowing. Nininger identified it as a meteorite and it became known as the **Coldwater (stone)** meteorite. Neither of the Coldwater meteorites turned out to be the meteorite that he witnessed, but they were twice as many meteorites as had been recorded in Kansas in the previous fifteen years. These demonstrated to him that there were many more meteorites out there to be found.

Nininger could not get funding from the College for this endeavor, and the university had no available funds. He decided that he would need to be self-supporting in his quest for meteorites. With the concurrence of the McPherson College President, he reduced his teaching schedule. He began to offer lectures for $10 once or twice a week throughout Kansas. Lectures at least paid his travel expenses but, more important to him, they were getting the word out about meteorites in the hope that they would result in new finds.

1927 Kansas Meteorite Farm

Ten tons (9.07 mt) of meteorites fell to earth about 20,000 years ago near what is now Haviland, Kansas (Figure 2.2). When the fall was first discovered in 1882, the closest town was the Brenham post office, so it became known as the **Brenham** meteorite. It is a rare, stony-iron meteorite (Figure 2.3).

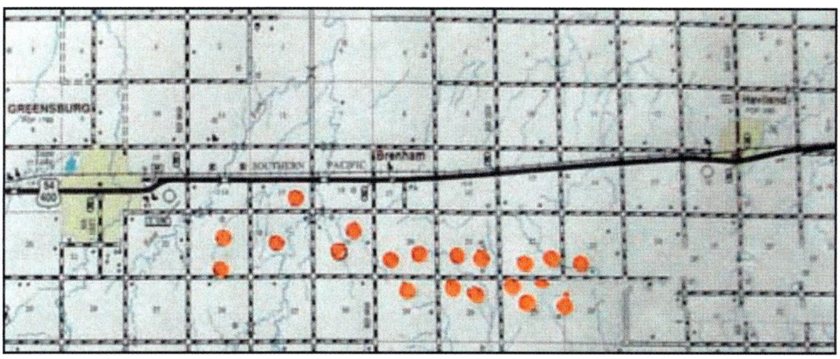

Figure 2.2 The Brenham meteorite scattered fragments as it descended. Image from Kansas Meteorite Museum.

The meteor that dropped the Brenham meteorites is estimated to have been 500 tons (453.6 mt). As it made its fiery entry through the atmosphere most of the material was vaporized but at least ten tons fell to the earth. The meteorites were identified by Native Americans as being of great interest. They were revered objects. Pieces of the Brenham meteorite were found in the burial mounds and on altars of the Hopewell Tradition in Ohio and Illinois (100 BCE to 500 CE). These were sometimes referred to as the Hopewell Mounds meteorites.

Figure 2.3 Slice of the Brenham stony-iron (pallasite) meteorite. Image courtesy of Geoffrey Notkin, Aerolite.org.

When the Brenham area was homesteaded, locals knew of these "iron rocks" and considered them unusual but useful. They were used as weights for rain barrel covers, corner stones for chicken houses, ballast for dugout roofs, or to hold down fence lines. During a visit to the area, local farmer Frank Kimberly told Nininger that he would plow them up regularly. His wife Mary thought they were meteorites and would pick them up to put in a pile. She wrote to an area geologist who came out to look at her collection. He was amazed at her hoard and paid her several hundred dollars for over a half-ton (453.6 kg) of material. Other scientists followed and the area became known as the "Kansas Meteorite Farm."

Nininger visited the Kimberlys several times, buying meteorite material on each visit. In 1927 he purchased a mass weighing 465 lb (210.9 kg). Over the years he purchased 1,200 lb (544.3 kg) of meteorites for his collection. Although he sold most specimens to fund other purchases

Figure 2.4. Nininger and son Bob with Brenham meteorites. Kimberly family watches from the house porch. Courtesy of the American Meteorite Laboratory Photo Archives, Collections Research for Museums, Denver, CO

and projects, his 1950 collection catalog lists 377 individual specimens at 374 lb (169.6 kg), plus a "box of fragments."

Figure 2.5. Haviland Crater before excavation. Dr. Clyde Fisher, head of the Hayden Planetarium, stands at the edge. Courtesy of the American Meteorite Laboratory Photo Archives, Collections Research for Museums, Denver, CO.

1929 Haviland Crater

In early 1929, during one of his visits to the Kimberlys, Frank was reminiscing about the sixty-eight pound (30.8 kg) meteorite he found in a "buffalo wallow." Nininger wanted to see this "wallow." The wallow was thirty-six by fifty-five feet (11 m x 16.8 m) in the form of a ellipse. It was said that it had been a lot deeper. Nininger noticed that this depression was surrounded by a definite rim, higher than the surrounding landscape. Frank advised that it held water longer than any of the other depressions on the farm and was employed as a watering place for stock during the early years of their residence there. Nininger said he was "elated at what I had found" and thought it was a meteorite crater.

Excavation of the wallow would be put off for several years while Nininger obtained funding. The excavation was also delayed because of Nininger's move to Denver to work at the Colorado Museum of Natural History (discussed in Chapter 1).

1929 Trip to Mexico

While studying meteorites he came across the *Catalogue of Meteorites of Mexico* published in 1889.[3] He recognized that there had been no more recent writings on meteorites in Mexico since that publication. Of course one of the reasons for this was probably the Mexican Revolution from

Figure 2.6. Vintage postcard of Instituto Geologico, Mexico City. Image available on the Internet and included in accordance with Title 17 U.S.C. Section 107.

1910 to 1920. Before that, there had been a body of reports on meteorites in the country. At the time, there was a greater incidence of meteorites in Mexico than in the United States. This may have been due to a more vigorous interest in meteorites by the Mexican government that declared that all meteorites were government property. In fact, Mexico City held five of the greatest meteorites found in the entire world at the time.

With the lack of new finds in Kansas, Nininger decided that a trip to Mexico might result in obtaining specimens that he could use to exchange for meteorites of more interest to him back in the United States, but the trip also presented the possibility of making some new discoveries.

Nininger had visited Mexico in 1928 as part of the Student Trip (Chapter 1). He began to plan a return visit in 1929. He approached one of his students, Alex Richards, who was with him on the previous Mexico visit, and asked him to accompany him. Alex agreed and plans firmed up. When they departed for Mexico City, they took a selection of meteorites that he had purchased for exchange, and a "great cargo" of food, camera equipment and other items. Over rough gravel roads, they would average forty miles (64.4 km) a day. They had been warned that there were areas of banditry, which they tried to avoid, but were often challenged by heavily armed men.

On their arrival in Mexico City they visited the *Instituto Geologico* (Figure 2.6). Asking about meteorites, they were directed to Dr. Frederick Mullerried, a German geologist who had traveled all over Mexico. He told them that no one was studying meteorites in Mexico at the time and there were a number of meteorite falls that had not been classified. The museum welcomed their help and they began examining the museum's collection. Some specimens were not labeled and some were mislabeled. For several weeks, Nininger and Alex worked in the museum cutting into samples to identify, polish, etch and label them. This experience honed their skills in meteorite cutting and identification.

Nininger was fascinated with the story of the **Toluca** or Xiquipilco meteorite found in 1776. Locals used some of the material to fashion into tools, but when scientists began to visit the site in 1824, fragments became items for sale. It was estimated that 21 tons (19 mt) had been sold by 1906. Nininger visited the location with Dr. Mullerried. They bought 700 lb (317.5 kg) of meteorites. He was also able to acquire additional specimens from various sites (Figure 2.7). A total of 900 lb (408.2 kg) of meteorites were shipped back to McPherson.

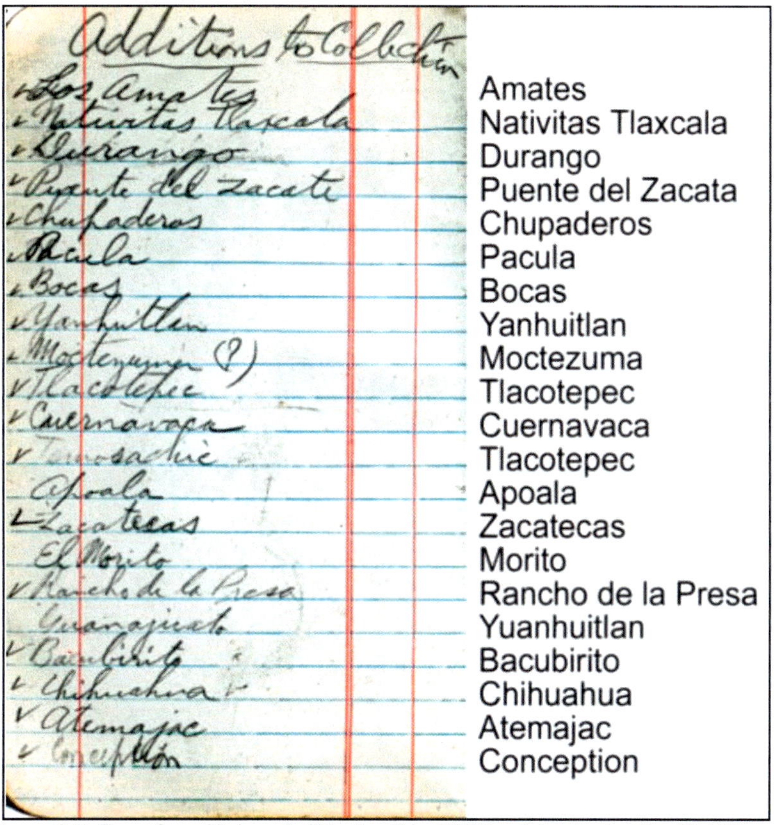

Figure 2.7. Page from Nininger's "Mexico Trip 1929" notebook showing "Additions to Collection" with text added for clarification. All items are shown in the 1950 Nininger Collection catalog. Courtesy ASU Buseck Center for Meteorite Studies.

1931 Huizopa Meteorite and Dean Gillespie

While at the *Instituto Geologia* in 1929, Nininger and Alex helped with some specimens that were not labeled or were mislabeled. Among those were small masses of the Huizopa Meteorite. This meteorite was found in 1907 in a ruin near Huizopa, Chihuahua, Mexico. When found, there was a main mass and four smaller masses. The main mass weighed 258 lbs (117 kg). The location of the main mass, however, was lost.

When the Niningers moved to Denver, their house was a few blocks from White Motor Trucks. In 1931, Nininger met the owner, Dean Gillespie, who was an amateur meteorite collector. When they met, they discussed meteorites and a close friendship began. Dean Gillespie was a wealthy businessman who had the White Motor Truck dealership.

Figure 2.8. 1931 advertisement for White Trucks Image available on the Internet and included in accordance with Title 17 U.S.C. Section 107.

Learning that Nininger was operating "on a shoestring," he offered him the opportunity to deliver his trucks from the dealership in Denver to locations throughout the West. He was paid $4 a day plus expenses as a part-time driver. He was allowed to visit museums, study collections and make exchanges with private collectors en route, as long as it was on his own time and expense. This arrangement lasted for about ten years.

When finishing a delivery to El Paso, Texas in 1931, Nininger decided to take a plane to Chihuahua City, Mexico, on the chance of finding the main mass of the Huizopa meteorite. Knowing that the smaller masses had been mislabeled, he always held the belief that the main mass was simply mislabeled and that he might be able to find it.

Nininger visited the State Mines Exhibit in the *Museo Historia Natural del Estado Chihuahua* in Chihuahua City. There was a pile of unlabeled ore samples on display. Examining the pile he discovered the main mass of the Huizopa meteorite. Since this belonged to the government, he contacted the United States consul. Recognizing the benefit of having the meteorite properly identified and studied, the government allowed Nininger to take the meteorite back to Denver where he would cut it in half and return half to the Chihuahua Museum. From his half, he made several slices to sell to museums, from which he received "a couple thousand dollars" for his efforts.

1931 Canadian Meteorites

In the summer of 1931, the Nininger family took a trip to Saskatchewan to visit Harvey's sister. During the drive, he noticed several rock piles along fence lines and thought this might be a likely place to find meteorites. Following his usual custom, he visited the local newspaper, the *Saskatoon Star* and suggested an article about meteorites. The editor asked Nininger to write the article. The article explained what a meteorite was, how to recognize it, and directed them to chip off a small piece and send it to him to examine.

A few weeks later he received a sample from a farmer in Springwater, Saskatchewan. The farmer, Mr. Ward, explained that he had taken the sample to the local assayer who thought it to be iron that leaked out of a furnace and had no value. He explained that he had found the parent mass, at about 44 lb (20 kg), a few years earlier. Nininger asked him to send the mass, which he did in exchange for payment. Ward keep looking and eventually found about a dozen pieces weighing about 200 lb (90.7 kg) in total. The Nininger 1950 Collection catalog lists the **Springwater** main mass (after slicing), and an end piece, at 84 and 15 lb (38 and 6.8 kg), respectively. The main mass of this meteorite is at the ASU Buseck Center for Meteorite Studies. Interestingly, in 2009, a piece weighing 117 lb (53 kg) was unearthed and now resides at the Royal Ontario Museum.

Several months later, Nininger received a small sample from Mr. A. D. Ebner, a farmer near **Bruno**, Saskatchewan. He explained that he was hauling rocks from his field and noticed that one rock was so heavy that he had to use two hands to pick it up. This reminded him of the newspaper article. The mass was purchased by Nininger. In describing it, he emphasized the beautiful "flight-markings." A meteorite may fall "nose down" through the atmosphere or it may tumble. Meteorites that do not tumble can assume aerodynamic, streamlined shapes as material is removed by atmospheric friction. In the photo of the Bruno meteorite

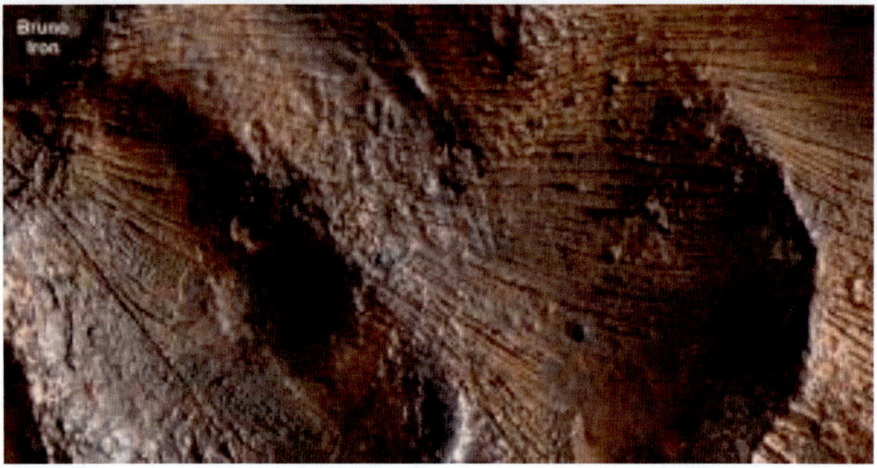

Figure 2.9 Close up image of the "flight markings" of the Bruno meteorite. Photo by L. Garvie, ASU Buseck Center for Meteorite Studies. Used with permission.

(Figure 2.9), lines stream back from the nose of the meteorite. Some meteorites even show trails of melted droplets that start at one point and fan out over the surface of the meteorite.

1931 Eaton Copper Meteorite Controversy

In May 1943, Nininger published a paper in *Popular Astronomy*[4] entitled "The Eaton Colorado Meteorite: Introducing a New Type." The paper described a meteorite that he had acquired from W. H. Foster. Foster was gardening when an objected whizzed past his face and landed a few feet from where he was standing. Foster picked up the object and reported that it was so hot that it burned his fingers. Puzzled, he took the object across the street to show John Casey, the high school superintendent, who then showed it to Mr. Rhodes, the school janitor. Glen Mills, the school science teacher, also examined the object. It had a copper appearance but did not look like a bullet or a meteorite. They thought it might have dropped off of an airplane, but none were seen that day.

About three weeks after the discovery, Mills was attending a meeting that Nininger was also attending. He told Nininger about the object and asked him to investigate. Nininger visited Foster, along with Dr. Figgins of the Colorado Museum of Natural History. The interviews from Nininger's notebook are shown in Figures 2.10 and 2.11. They agreed that it did not look like a meteorite but it did have the surface pitting and flight markings common with meteorites.

Foster loaned the item to Nininger (Figure 2.12) to study. Nininger

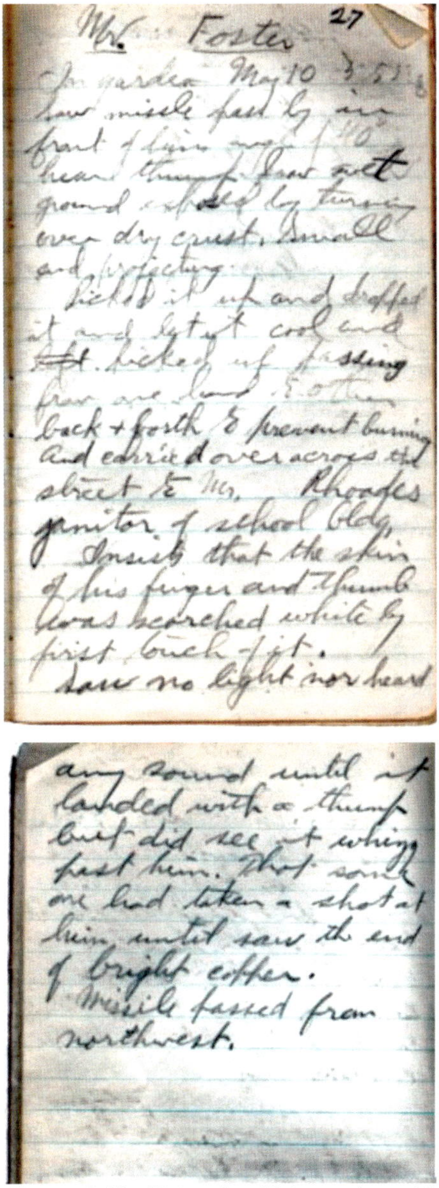
Mr. Foster 27
In garden May 10 3:55
saw missile pass by in
front of him [illegible] 140
heard thump [illegible]
ground [illegible]
over dry crust. Small
and [illegible]
Picked it up and dropped
it and let it cool and
[illegible] Picked up passing
from one hand to other
back + forth to prevent burning
and carried over across the
street to Mr. Rhodes
janitor of school bldg.
Insists that the skin
of his finger and thumb
was scorched while he
first touched it.
Saw no light nor heard
any sound until it
landed with a thump
but did see it whizz
past him. Thot some
one had taken a shot at
him until saw the end
of bright copper.
Missile passed from
northwest.

Figure 2.10. Notes by Nininger from his interview with Foster. Courtesy ASU Buseck Center for Meteorite Studies.

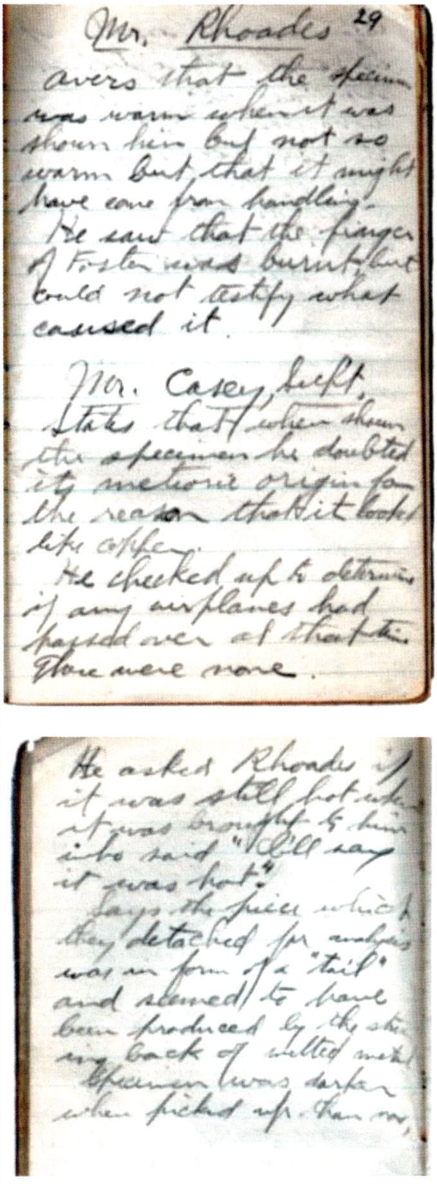
Mr. Rhodes 29
avers that the specimen
was warm when it was
shown him but not so
warm but that it might
have come from handling.
He saw that the finger
of Foster was burnt but
could not testify what
caused it.
Mr. Casey, Supt.
States that when shown
the specimen he doubted
its meteoric origin for
the reason that it looked
like copper.
He checked up to determine
if any airplanes had
passed over at that time.
There were none.
He asked Rhodes if
it was still hot when
it was brought to him
who said "I'll say
it was hot."
Says the piece which
they detached for analysis
was in form of a "tail"
and seemed to have
been produced by the [illegible]
ing back of melted metal.
Specimen was darker
when picked up than now,

Figure 2.11. Notes by Nininger from his interview with Rhodes and Casey. Courtesy ASU Buseck Center for Meteorite Studies.

eventually paid Foster $5 for the object. Knowing that this would be controversial, his report of the find in a 1943 article begins:

> *"In describing a new and radically different type of meteorite, I am fully aware of my obligation to all astronomers,*

geologists, and meteoriticists, as well as to the principle of accuracy. Had it not been for this awareness, the present announcement might have appeared some years earlier."

He goes on to acknowledge that "In presenting the account which is to follow in this paper, I am confidently expecting to be criticized."

Its original weight was only 0.09 oz (29.5 g), with an irregular shape. His hypothesis was that the specimen was a meteorite of an entirely new type (copper-zinc-lead). He was correct in his prediction as some exclaimed that it was a "Hoax." However, he had the courage to publish his finding and remark: "I am willing to have the matter rest in the lap of time."

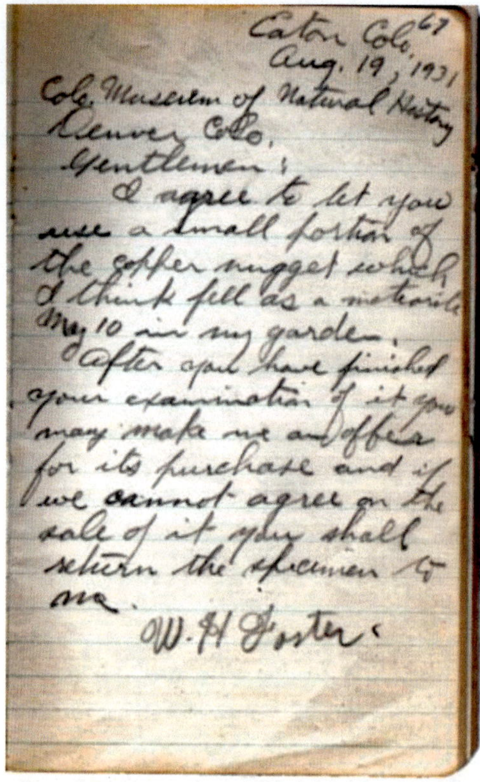

Eaton Colo. 67
Aug. 19, 1931
Colo. Museum of Natural History
Denver Colo.
Gentlemen:
I agree to let you use a small portion of the copper nugget which I think fell as a meteorite May 10 in my garden.
After you have finished your examination of it you may make me an offer for its purchase and if we cannot agree on the sale of it you shall return the specimen to me.
W. H. Foster.

Figure 2.12. Nininger notebook entry of the agreement between Foster and Nininger. Courtesy ASU Buseck Center for Meteorite Studies.

Copper occurs, in various degrees, as a minor component in all types of meteorites. Nininger's report, in announcing the existence of a new meteoritic type, found that the Eaton object consisted of copper, zinc and lead.

A 1973 study of Eaton[5] showed that it is a cooper-zinc alloy containing 66% copper and 33% zinc, with less than 1% nickel. The study also compared Eaton's copper content with different groups of meteorites and terrestrial rocks that contained copper. The main conclusion of the study was that Eaton is not a meteorite on the basis of its much higher content of zinc and lower content of nickel in comparison to the copper in other meteoritic bodies. The study concluded that Eaton is an artificial yellow brass.

However, later studies suggest that this may not be the final answer. Other meteorites have been studied that contain different copper alloys. In addition, the Apollo lunar mission returned lunar material that has a similar copper-zinc composition to Eaton.[6] It appears that the "lap of time" is not finished with the Eaton meteorite.

1931 Holbrook Meteorite

> *Friday evening about six-thirty a meteor, or some other body of a like nature, passed over Holbrook going almost due east at a rate of speed that would make a swift-moving express train seem as though it were standing dead still. The noise it created was very loud and lasted for at least a half a minute and sounded somewhat like distant thunder or the booming of a cannon in the distance.*

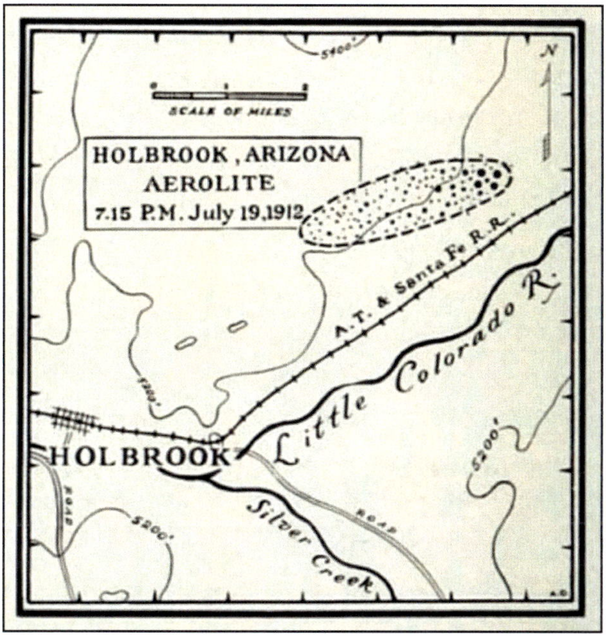

Figure 2.13. Map in the 1933 American Museum of Natural History Field Guide[8] of the Holbrook Shower. Image available on the Internet and included in accordance with Title 17 USC Sect. 107.

Above is from an Arizona newspaper[7] on the arrival of what became known as the **Holbrook** Meteorite. In the late afternoon of July 19, 1912, a spectacular meteorite shower was seen over several states. The fall began with a loud blast followed by smaller explosions and a rumble. More than 14,000 meteorites fell near the Santa Fe Railway Aztec yard, six miles east of Holbrook, Arizona. The fall was witnessed by hundreds of people who quickly gathered fragments. The largest was 14.5 lb (6.6 kg) with others as small as a few ounces. A number of them had well developed fusion crusts. Fusion crust is a thin, melted surface layer of thermally transformed components of a meteorite. It forms when the meteoroids enter earth's atmosphere between 10 and 44 miles (15 and 70 km) per second.

The Niningers visited the site in 1931. On that and subsequent visits, they recovered a number of small fragments. The 1950 collection catalog shows this as one of the largest collections that included 80 individuals, 205 fragments, 400 small fragments, 960 small stones in "a glass jar," and a "small panel containing 237 small individuals and fragments."

1931 Beardsley Meteorite

On October 15, 1929, residents of Beardsley, Kansas, were awakened about 11:30 pm by the light of a fireball and "unusual thunderous sounds."[9] Nininger was still in Mexico. By the time he had heard of the fall, two recovered stones had been sold to a collector in Michigan. On their way back from Canada in 1931, he decided to visit the collector to see and discuss the finds. As they continued home, they stopped in Beardsley and talked to the farmer who had sold the two stones. On August 10, 1931, Nininger began a 60-day search for additional specimens. He was rewarded with six masses found by farmers. The specimens were either lying upon the ground or plowed up just a few inches from the surface.

All the stones were covered with the usual dark, slaty grey, fusion crust. The largest specimen collected was 20.5 lb (9.3 kg), which was one of 27 specimens totalling 27 lb (12.2 kg) in the 1950 collection catalog.

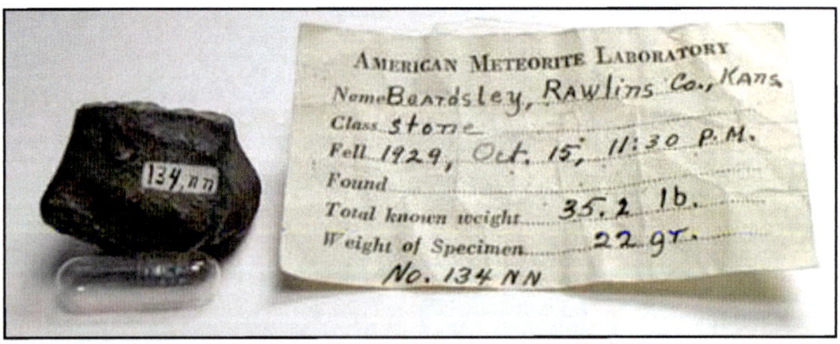

Figure 2.14. Most specimens from the Beardsley meteorite fall were small such as this specimen at 0.78 oz (22 gr). Photo courtesy of Martin Horejsi.

1932 Archie Meteorite

Nininger submitted a paper to *Popular Astronomy*[10] to describe his search for a meteorite that had been seen to fall on August 10, 1932. As described earlier, his method included visiting the area and interviewing as many people as possible who had seen the meteorite fall in hopes of finding fragments. He visited the area for two days in January 1933 and ten days in April 1933. He also gave five public addresses on meteorites. He estimated that he reached over 1,000 people in five towns in this way.

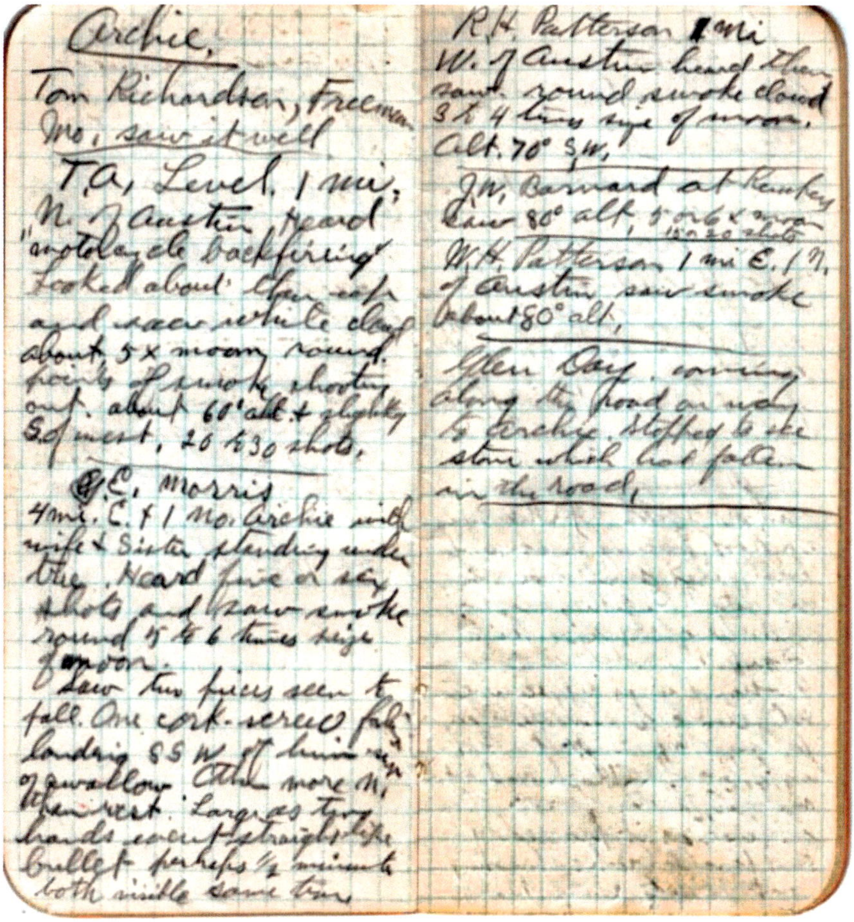
Archie,
Tom Richardson, Freeman Mo, saw it well
T.A. Level, 1 mi, N. of Austin heard "motorcycle backfiring" Looked about [illegible] and saw white cloud about 5 x moon round. [illegible] of smoke shooting out. about 60° alt + slightly S. of west, 20 to 30 shots,
G.E. Morris 4 mi. E. + 1 N. of Archie with wife + Sister standing under tree. Heard five or six shots and saw smoke round 5 to 6 times size of moon.
Saw two pieces seen to fall. One cork-screw fall landing SSW of him in swallow. Other more N. than west. Large as two hands went straight like bullet perhaps ½ minute both visible same time
R.H. Patterson 1 mi W. of Austin heard the same sound smoke cloud 3 to 4 times size of moon. Alt. 70° S.W.
J.W. Barnard at [illegible] saw 80° alt, 5 or 6 x moon 15 or 20 shots
W.H. Patterson 1 mi E. 1 N. of Austin saw smoke about 80° alt,
Glen Day, coming along the road on way to Archie, stopped to see stone which had fallen in the road,

Figure 2.15. Page from Nininger notebook on the Archie meteorite. Courtesy ASU Buseck Center for Meteorite Studies.

Figure 2.15 shows two pages from a notebook that Nininger labeled "Archie Mo." These pages detail interviews with seven of the sixteen individuals recorded in the notebook. While we know that he used this method of interviewing people, it is interesting to actually see his handwritten notes about what he was told. For example, T.A. Level, "heard a motorcycle backfiring." G.E. Morris "with wife and sister standing under a tree heard five or six shots and saw smoke round 5 to 6 times size of moon." On another page, he describes the lengthy account of Ollie E. Grant who "was walking along path on a slope heard a blast (made earth tremble) followed by 2 more blasts and smoke." He goes on to describe the sound "first a hissing like escaping steam . . . fireball then came, a low rumble like horse running on ground. This grew louder . . ."

Many of those interviewed were able to collect specimens. Nininger pointed out that this was unusual, but this was possible since it fell in a populated area. He mentions that one piece fell "about eighteen inches from a man's head." This investigation yielded seven pieces, the largest being about 8 lb (3.6 kg). He estimated the entire mass was 11.44 lb (5.2 kg). The 1950 Nininger Collection catalog lists one individual specimen at 0.87 lb (0.4 kg).

1933 Plainview Meteorite

In 1915, a Texan sent what he thought was a small stony meteorite that he found near Plainview, Texas, to the Smithsonian Institution. The Smithsonian determined that it was a new find. Up until that time only 22 meteorites had been found in the whole United States that had not been see to fall. The man was paid and was asked to try to search the area for more stones. Over the next two years of hunting, he collected and sent into the Smithsonian about a dozen stones at an average of 5 lb (2.3 kg) each. Dr. George Merrill of the Smithsonian wrote a paper in 1917 on the fall.[11] The Smithsonian concluded that the collecting was finished.

In 1928, Nininger wrote to Merrill about this find. He was interested in exploring the area further. Merrill gave Nininger the facts in a manner that Nininger described "like talking to me as a country school teacher might to a first-grader." Nininger then asked if any further searches of the area were going to be conducted. He was told the job was finished and that any further searches would, most likely, not yield further material. He then asked if there would be any objection if he tried his hand at searching the area. He was told there would be no objection but he offered little encouragement.

Working on other projects, he set this aside until 1933. He and his brother John were returning from a search in Mexico for a possible 50-ton (45.4 mt) meteorite, which turned out to be false, as well as tracking down a dozen other stories that turned out poorly. They neared Plainview, Texas, at sundown. He told his brother that they would stay there that night but, before retiring, he would go out into the countryside and find a meteorite. His brother looked at Harvey with real alarm after their thankless journey into Mexico with terrible food and water, insects, snakes and scorpions that had made their adventure into more of a nightmare. Plus, they were returning broke.

After getting a hotel room, Nininger and his brother drove into the country side and began knocking on farmers' doors just as the men were returning from the field. There were no results after talking to people at three farm houses, but they were told that it was possible that some rocks had been seen like that. They then stopped at a fourth farm house. The family was eating its evening meal by lamp light. They treated Nininger coolly and with puzzlement. When Nininger felt the impatience was giving way to hostility, he manage to tell them he would pay for such a stone at a dollar a pound. With the offer of money, the father jumped to his feet, took the lantern and went out into the courtyard with his eight-year-old daughter. They returned after a couple minutes with an 8 lb (3.6 kg) meteorite, asking Nininger if this is what he was after. Nininger purchased the meteorite and asked if there might be more. The man assured him there were but would have to be looked for in the morning. Nininger went back to the car, laying the meteorite in his brother's lap.

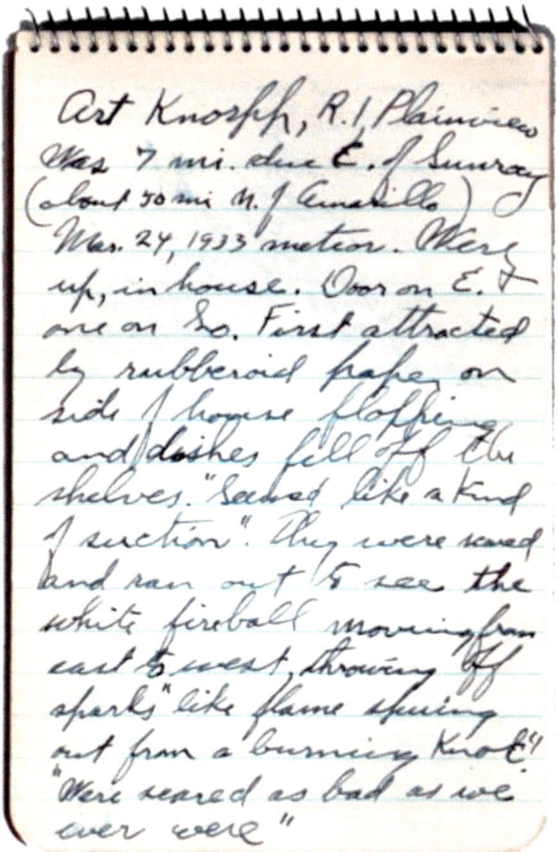

Figure 2.16. Plainview Meteorite interview from the Nininger notebook. Courtesy of ASU Buseck Center for Meteorite Studies.

In the morning they continued their search and by evening had acquired twenty-six meteorites, totaling 152 pounds (69 kg), paying a dollar a pound. Another man approached Nininger with a fist-sized meteorite that he used to throw at one of his bulls if it "goes on a rampage. I only need to hit him once with this, somewhere around the head, and he quiets right down." He purchased that one as well.

Going to a neighboring farm he asked the woman who answered the door if she had any stones like the one he was showing her. She was not friendly until he offered to pay for them. She then went inside and returned with two meteorites that were used as "weights in the pork barrel." She then brought him three more. Looking at the check she said "that's more than I got off the farm this year." Continuing down the line of farms they found meteorites at many farms with farmers grateful for the cash during that dust-blown depression year of 1933.

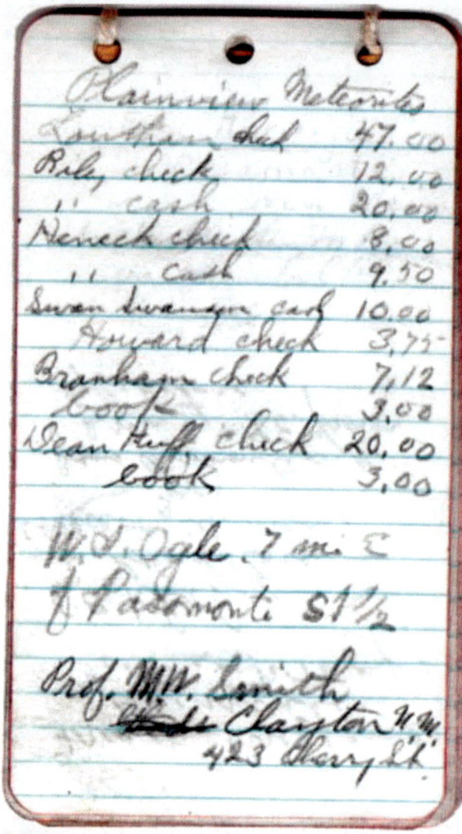

Figure 2.17. Plainview Meteorite purchases from the Nininger notebook. Courtesy of ASU Buseck Center for Meteorite Studies.

During repeated trips to the Plainview area, Nininger had collected 900 meteorites by 1948 in an area that Merrill thought would yield nothing. After selling or exchanging

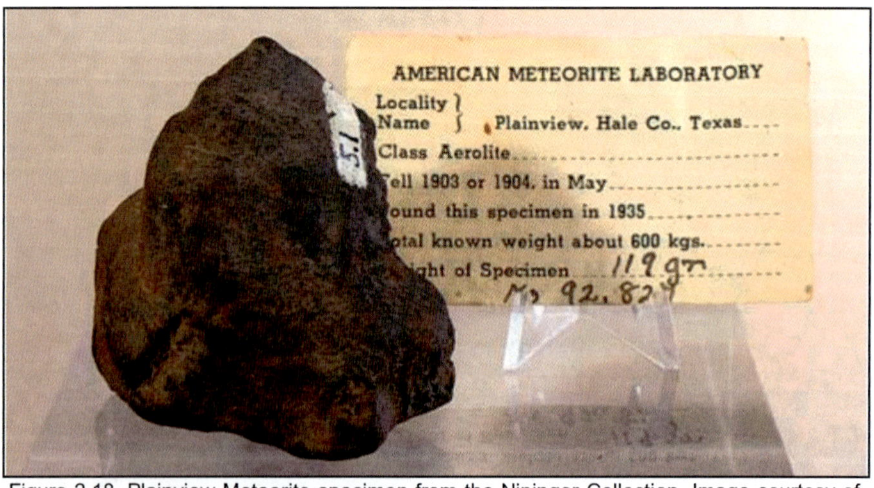

Figure 2.18. Plainview Meteorite specimen from the Nininger Collection. Image courtesy of Geoffrey Notkin, Aerolite.org, with permission.

some of these meteorites, his 1950 collection catalog showed the Plainview meteorite collection as the largest with 244 specimens totalling 434 lb (197 kg), with the largest weighing 14.2 lb (6.4 kg) and the smallest at 0.03 oz (.85 gm) Interestingly, one of Nininger's Plainview meteorite specimens, weighing 3.66 lb (1.66 kg) was sold in 2020 by Christie's auction house for $7,500,[12] a big difference from the $1 a pound he paid.

1933 Return to Haviland Crater

While working at the Denver museum, Nininger was anxious to return to the Kimberly farm's "buffalo wallow" to prove that it was formed by a meteorite. Dr. Figgins, Director of the Colorado Museum of Natural History, agreed to fund the expedition. Returning in 1933, Nininger hired two teams of horses and two road scrapers. Several cuts were made in the wallow, excavating and examining the soil of each cut. Each cut yielded many meteorites ranging in size from grapeseed-like to 15 lb (6.8 kg) When enough of the crater fill was removed with scrapers, crews worked with hand shovels.

Figure 2.19. Haviland excavation crew. Courtesy of the American Meteorite Laboratory Photo Archives, Collections Research for Museums, Denver, CO.

The discovery and excavation of the **Haviland Crater** was the first example of a meteorite crater being excavated so as to reveal the distribution of meteorites within, and suggesting the manner of its formation.

Figure 2.20 is a page from Nininger's notes on the excavation. Other notes detail finding many meteorites of various sizes at different depths, including "5 sacks" at 5 ft (1.5 m). A hand-drawing in the notes shows three specific discoveries:

> "Took about 200 lbs from depth of 6 ft. 2 in."
> "At 7 ft. 6 in. we were still getting chunks of 20 lbs and under."
> "100 lb mass at 9 ft. 9 in."

In a 1933 paper, Nininger and Figgins[13] suggested that the pit was formed by the impact of several smaller meteorites in close proximity (known as a "swarm"). The meteorite fragments were found to occur in a funnel shaped formation, the smaller fragments peripherally toward the surface of the ground, and the larger ones nearer the center at the bottom of the funnel.

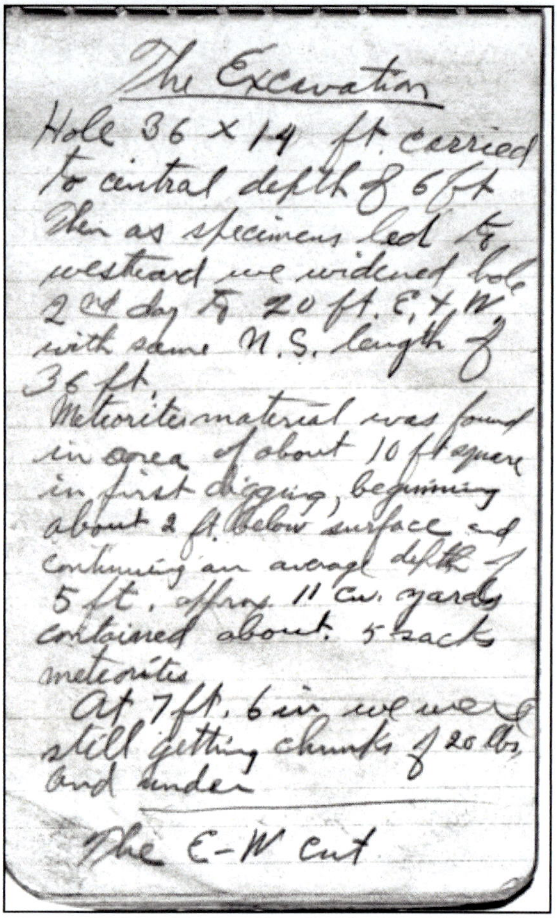
The Excavation
Hole 36 x 14 ft. carried to central depth of 6 ft then as specimens led to westward we widened hole 2nd day to 20 ft. E. & W. with same N.S. length of 36 ft.
Meteorite material was found in area of about 10 ft square in first digging, beginning about 2 ft. below surface and continuing an average depth of 5 ft. approx. 11 cu. yards contained about 5 sacks meteorites
At 7 ft. 6 in we were still getting chunks of 20 lbs. and under
The E-W cut

Figure 2.20. Page from Nininger notebook on Haviland excavation. Courtesy ASU Buseck Center for Meteorite Studies.

This arrangement was interpreted as evidence of an explosion of the meteorite, or of the air trapped between it and the soil at the instant of, or just before, the impact. Particles in the crater consisted of microscopic grains of oxidized meteoritic material chemically, and mineralogically, identical to the larger oxide fragments found.

At the time of Nininger's excavation, the Haviland Crater was one of only three craters that were authenticated by the presence of meteorites. Nininger's collection catalog included a listing of 377 specimens from the excavation, and purchases from, the Kimberly family.

1932-1935 Odessa Meteorite

The first mention of the Odessa, Texas, meteorite crater was made in a 1926 report[14] of finding an iron meteorite fragment in a shallow depression. Another person interested in the Odessa Crater was Daniel Moreau Barringer, a highly successful mining entrepreneur who had filed

Figure 2.21. Excavation in progress within the Haviland Crater. Courtesy of the American Meteorite Laboratory Photo Archives, Collections Research for Museums, Denver, CO.

Figure 2.22. Profile of Haviland Crater excavation showing fragments in the wall. Courtesy of the American Meteorite Laboratory Photo Archives, Collections Research for Museums, Denver.

claims for Meteor Crater in Arizona (discussed at length in Chapter 3). Barringer was having difficulties convincing the scientific community that Meteor Crater was, in fact, created by a meteorite. Upon reading about the Odessa crater in 1926, he thought it might help his argument. His son, Daniel Moreau (Reau) Barringer, Jr., visited the site and was convinced it was caused by a meteorite. Barringer Sr. suggested, in a letter to his son dated June 25, 1926, that the meteorite that created Odessa Crater "may easily have been a traveling companion" of the Arizona meteorite, forming at the same time. Reau Barringer wrote a paper[15] of his findings but publication was held until 1928. Thinking there might be a meteorite in the crater worth extracting, Barringer wanted time to purchase the forty-acres containing the crater before its publication. However, since the land also contained oil and gas potential, the owners refused to sell.

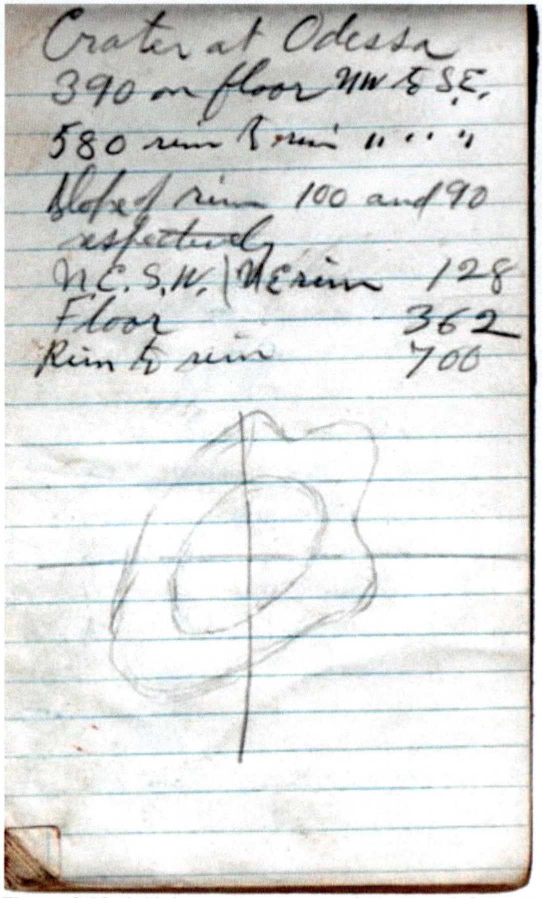

Figure 2.23. Initial measurement and sketch of Odessa Crater. Courtesy ASU Buseck Center for Meteorite Studies.

The earlier excavation at Haviland Crater increased Nininger's interest in the shallow depression at Odessa. The Haviland crater was relatively small in comparison. Figure 2.23 is a page from Nininger's notebook on the site where he measures and sketches the crater. Nininger first visited the site in 1932 with an "electro-magnet" (Figure 2.24) which was attached to a battery. He covered the depression and collected about 1,500 fragments totalling eight pounds (3.6 kg). He believed that this was the first "electro-magnetic rake" ever used for meteorite collecting.

He returned with his son Bob in 1933. This time they had a new invention called a "magnetic balance," the forerunner of today's metal detector. This instrument operated according to the principle of a pendulum or beam balance to measure the magnetic susceptibility of various objects.

Within the first fifteen minutes of scanning the surface with the instrument (Figure 2.25), Bob stopped and declared that he had found something. Nininger dug just under the surface and found a one pound meteorite. The next morning, Nininger walked around the depression and spotted a meteorite poking out from the rim. He dug out an eight and a half pound meteorite, the largest found up to that time at Odessa.

Figure 2.24. Electro-magnet pulled behind car. Courtesy of the American Meteorite Laboratory Photo Archives, Collections Research for Museums, Denver, CO.

Figure 2.25. Bob Nininger examines field with an electro-magnet. Courtesy of the American Meteorite Laboratory Photo Archives, Collections Research for Museums, Denver, CO.

They continued to use the magnetic balance instrument and each time Bob indicated a spot to dig, they found a meteorite with the deepest at seven inches below the surface. A total of twenty-seven meteorites weighing 34 lb (15.4 kg) were found with the device. He first reported this activity in a 1934 issue of *Popular Astronomy.*[16] In this issue he stated: "This writer desires to report certain facts concerning this crater which were gathered by him in May 1933, and to go on record as one who believes that there is not the slightest doubt that the Odessa, Texas, crater was produced by the impact of a large meteorite."

1933 Melrose Gold Meteorite

Another example of Nininger's "process" ocurred while in Clovis, New Mexico. Nininger visited the editor of the *Curry County News*. The editor then wrote a story about Nininger and meteorites that appeared the following Thursday. On Saturday, the editor was visited by rancher who lived near Melrose, New Mexico, with a 68.5 lb (31.1 kg) rock that he kept hitting with his plow. He finally dug it up and was using it as a weight. Nininger had left the area but upon describing it to him he sent the rancher a check and said he would pick it up on his next trip. The editor wrote a follow-up story about the sale and more meteorites showed up. Nininger asked the chief chemist for the Inspiration mine and smelter at the Anaconda Copper Company in Miami, Arizona, to analyze the **Melrose** stone. The most striking characteristic was the discovery of gold among its mineralogical features. As this was the first instance of finding gold in a meteorite, the sample was also sent to the American Smelting and Refining Company in Denver for analysis. Their test agreed that very small traces of gold were present within the meteorite. After removing sections for testing and sale, the Melrose meteorite in Nininger's 1950 collection catalog was reported at 47.7 lb (21.3kg).

When Nininger published the result of this find,[17] it made headlines "all over the world." He reports that this complicated his work in that some people thought his interest in meteorites was to seek precious metals. Still others demanded much higher prices for their meteorites, and still others were cutting up meteorites to find any gold content. He said he had to deal with this misconception for several years.

1935 Morland Meteorite

Nininger continued his practice of giving lectures throughout Kansas. In 1933 he gave a lecture at a school in Sharon Springs, Kansas. After this talk he met a Mr. Dollarhide. There was no further contact with him until

1935 when Dollarhide sent Nininger a chip from what he thought was a meteorite that he found near the village of Morland that he estimated was several hundred pounds.

Nininger visited him at the site. Dollarhide explained that about 1890 he had broken his plow when he hit this mass that he described as "big as a wash tub in a field where I've never even seen a pebble before." When Nininger saw the mass he said that it "set my nerves tingling." The mass weighed over 600 lb (272.2 kg). It had broken into three pieces but could be fitted back together. Had Dollarhide not been at Nininger's talk in 1933, the meteorite may never have been discovered. **Morland** remains one of the most massive of over 5,000 meteorites in its class of chondrites. The main mass at 628 lb (284.9 kg) was sold to the British Museum of Natural History in 1958. An 11 lb (5 kg) specimen is at the ASU Buseck Center for Meteorite Studies.

Figure 2.26. Nininger photo of Morland meteorite loaded for transport. Courtesy of the American Meteorite Laboratory Photo Archives, Collections Research for Museums, Denver, CO.

1935 Hugoton Meteorite

Nininger and son Bob were returning from a trip to New Mexico when they stopped in Hugoton, Kansas. They stayed the night with former students. He arranged for a lecture at the local high school where the principal was also a former student. At the end of the talk a student came up to Nininger and told him about a stone that resembled the specimen he brought.

They visited the farm and found the spot where the father remembered hitting it with a plow. They soon found meteorite fragments scattered in the area, probably broken from plowing. They started to dig where the concentration of fragments was the highest. When they struck rock they began to dig with their hands so as to not damage the weathered meteorite.

By noon, they found a 749 lb (340 kg) stony meteorite, along with 21 lb (9.5 kg) of fragments (Figure 2.27). This mass, from a cornfield in 1935, represented the discovery of the second largest known stony meteorite in the world, at the time. As of 2016, **Hugoton** was the 5th most massive of over 8,000 stony meteorites of its class.

Figure 2.27. Nininger with the Hugoton meteorite. Courtesy of the American Meteorite Laboratory Photo Archives, Collections Research for Museums, Denver, CO.

1933-1936 New Mexico Meteorites

While in Clovis, New Mexico, Nininger received reports of a fireball falling in the vicinity of Pasamonte, New Mexico in 1933. Charles Brown, a cowboy in Clayton, New Mexico, noticed that the sky had lit up to "noontime brilliance." He grabbed his camera and caught a picture of an approaching meteorite (Figure 2.28).

Nininger, and son Bob, went to the area and started to interview anyone who saw the fall. Reports varied over the next four months as they tried to narrow down where it landed. On several trips to the Pasamonte area, he made a point to talk to the ranch hands on recognizing meteorites in the field. While visiting one of the ranches he was told about the Lanfor farm. As he drove there he noticed several rocks along the way, and then a pile of rocks near the farm house. The rocks were from the **Pasamonte** meteor. He purchased the rocks. His 1950 collection catalog records 31 fragments.

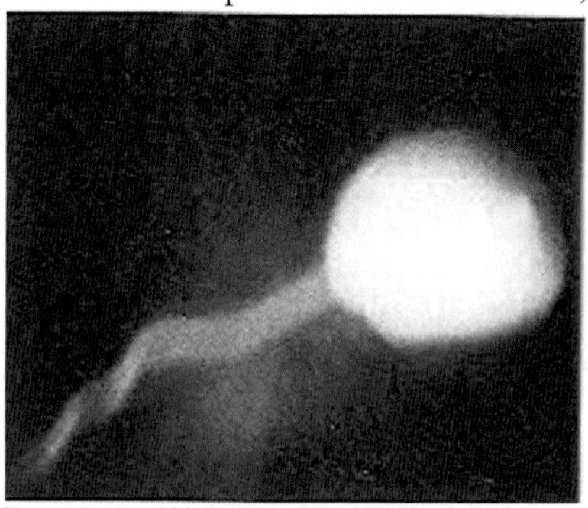

Figure 2.28. 1933 Pasamonte meteorite image taken by Charles Brown. Image available on the Internet and included in accordance with Title 17 USC Section 107.

Nininger's former student, Alex Richards, continued to work for Nininger. In addition to lectures and newspaper articles, they would also visit promising locations. In 1936, Nininger asked Richards to go to Gladstone, New Mexico, for three weeks and set up a display of meteorites in the country store and "try to contact every farmer." He assumed there had to be some meteorites in the area since none had previously been reported and the land was under cultivation. Near the end of his stay a rancher approached Alex to say that he had "one as big as a coal bucket" lying against the barn ever since he plowed it up four years earlier. Alex visited the farm and purchased the 128 lb (58 kg) meteorite. During his stay, he purchased four others from three different falls, namely, **Gladstone**, **Farley** and **Ute Creek**, all within six miles of the Gladstone post office.

1938 Camp Verde Meteorite

In 1928, George Dawson, "an amateur collector of Indian relics" from Phoenix, uncovered a stone-lined cist in "a ruin along Clear Creek east of Camp Verde."[18] It was described as "a little pocket in the earth walled and covered over with flat rocks." Dawson recounted that he thought it was a child's grave. It was covered in about 15-18 in (38-46 cm) of loose dirt. In the cyst he found a "feather cloth" and inside the cloth was a large "oak leaf-shaped" object. In 1938, Dawson wrote to Nininger saying: "I have been told you buy meteorites. I have one of unusual appearance in the shape of, and looks like, it may be a giant arrowhead. Would you buy this?"[19]

Figure 2.29. Geprge Dawson is standing in the depression where the Camp Verde meteorite was found. Courtesy of the American Meteorite Laboratory Photo Archives, Collections Research for Museums, Denver, CO.

Dawson had not mentioned this find before because he was concerned about the Antiquity Act that prohibited digging in archaeological sites on federal land without a permit. The site was on the Coconino National Forest. Nininger incorrectly told him that it was not a concern since "the law had been adopted some seven years after he made his discovery." The Antiquities Act was the first U.S. law to provide legal protection of cultural and natural resources of historic or scientific interest on Federal lands. President Theodore Roosevelt signed the Antiquities Act on June 8, 1906. The Act set an important precedent that asserted a broad public interest in the preservation of these resources on Federal lands. Legally, the Camp Verde meteorite should have been sent to the Smithsonian's National Collection of Meteorites.

Nininger wrote back that he would stop by to look at it next time he was in Phoenix. When he did visit and examine the meteorite, he wanted to see where it was found. Dawson drove them to the Camp Verde site. Nininger recounted how they "reached the crumbled walls of a small room, in the corner of which there was a slight depression and several flat stones protruding from the drifted dust and debris."

After visiting the site, Nininger purchased the meteorite for $75. The meteorite weighs 135 lb (61.2 kg) and is about 23.6 in (60 cm) long, 11.8 in (30 cm) wide (Figure 2.30). It has a thickness of 5.5 in (14 cm) at the base. Nininger gathered thirty-nine pottery sherds while at the site and sent them to archaeologist John Cotter at the nearby Tuzigoot National Monument. Cotter advised that the pottery sherds were dated from the 10th and 11th century, which would put them during the time of the Hisatsinom (Hopi ancestors).

Figure 2.30. Camp Verde meteorite located at ASU Buseck Center for Meteor Studies.

The location of the dwelling where the meteorite was discovered had been the subject of speculation for many years. It was described as being "near Road Forks, 6 miles below Verde" by Cotter. Camp Verde was referred to as "Verde" for many years. Dawson only referred to the location as "east of Camp Verde along Clear Creek." Nininger identified the location as "some 9 miles east of the Camp Verde post office, a small mesa stands about 100 feet above one of the boulder-strewn washes which are tributary to the Verde River," but the roads are different today for tracking mileage and the post office is in a different location from 75 years ago. Peter J. Pilles, Jr., Coconino National Forest Archaeologist (pers. com. 2005), suspected that it was in a ruin referred to as the "Garden Site," later renamed the John Heath Ruin. A photo was found of Dawson standing in the recessed location of the meteorite (Figure 2.29). Based on the surrounding landscape in the photo, it was confirmed that the John Heath Ruin was the actual location of the discovery (Figure 2.32). Remnants remain of the cist in which it was located.

Studies by the Buseck Center for Meteorite Studies confirmed that the **Camp Verde** meteorite is a coarse octahedrite from the Canyon Diablo fall, east of Flagstaff, that created Meteor Crater. The way in which it was wrapped, in a feather cloth, is reminiscent of the treatment of the Casas Grandes, Mexico, meteorite[20] that was wrapped in a linen or mummy cloth. Placing it in a stone-lined cist was also similar to how the Winona Meteorite[21] was found in a Hisatsinom ancient dwelling east of Flagstaff. Based on the similarity of treatment (tomb-like structure or stone boxes, and wrapping) it has been suggested that these objects were regarded as the remains of a deceased celestial being falling to the earth. In looking for the feather blanket in which it was wrapped, a note was located from Dawson indicating that he had cut it into 3 inch squares to give to his customers at his Phoenix store.

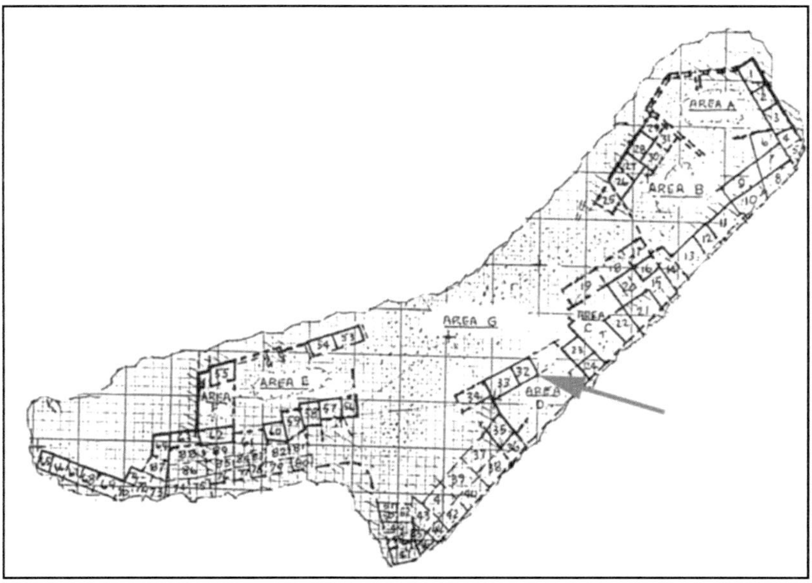

Figure 2.31. The John Heath Ruin, on a boot-shaped mesa, was the location of the Camp Verde meteorite, indicated by an arrow. Drawing by Shelby Coody and provided by the Verde Valley Archaeology Center.

1937-1938 Miami and Potter Meteorites

During 1937, Nininger and his team had recovered thirty-one previously unknown meteorite falls. None were seen to fall but his many lectures, newspaper interviews, and magazine articles (and probably the idea of getting $1 a pound) had stirred the population to look for and report meteorite finds.

For example, on a trip through Miami, Texas, Nininger was visiting a friend who told him about a farmer who told him about a stone that could be a meteorite. He visited the farmer who remembered that such a rock was used as a weight in ditches and washes to hold the fence wire. With a little digging, Nininger found the 127 lb (57.6 kg) **Miami** meteorite.

On another occasion, Nininger was in Colorado and stopped at a diner. When he sat down he placed a small stony meteorite by his water glass. A short time later, a customer came in and sat next to him. The man picked up the stone to examine it. Nininger told him what it was. The man told Nininger "my brother-in-law in Nebraska may have one of those things in his yard. It's a big thing." He asked the man to send his brother-in-law his leaflet on how to recognize a meteorite. A few weeks later he received a small meteor fragment. Nininger went to **Potter**, Nebraska where the farmer had a 450 lb (204 kg) meteorite in his yard that he had plowed up a few years earlier. He also had another fragment, 150 lb (68 kg), from the site where he had plowed up the first specimen. Nininger paid for and collected both specimens. The main mass is now at the Field Museum of Natural History in Chicago. Another large mass is at the ASU Buseck Center for Meteorite Studies.

Collecting continued in 1938, including the **Goodland** meteorite from Kansas, the **Cope** meteorite from Colorado, and **Aurora** meteorite from New Mexico.

1938 Goose Lake Meteorite

The **Goose Lake** meteorite, a 2,573 lb (1,167 kg) iron, was found on October 13, 1938, lying on ancient lava beds about two miles west of Goose Lake in the Modoc National Forest in northeast California, and about 1¼ miles south of the Oregon border, by three deer hunters.[22] The hunters thought it might be a meteorite so managed to take a small sample from the mass. Clarence Schmidt, one of the discoverers, took the sample and a photo to Dinsmore Alter, the Director of the Griffith Observatory

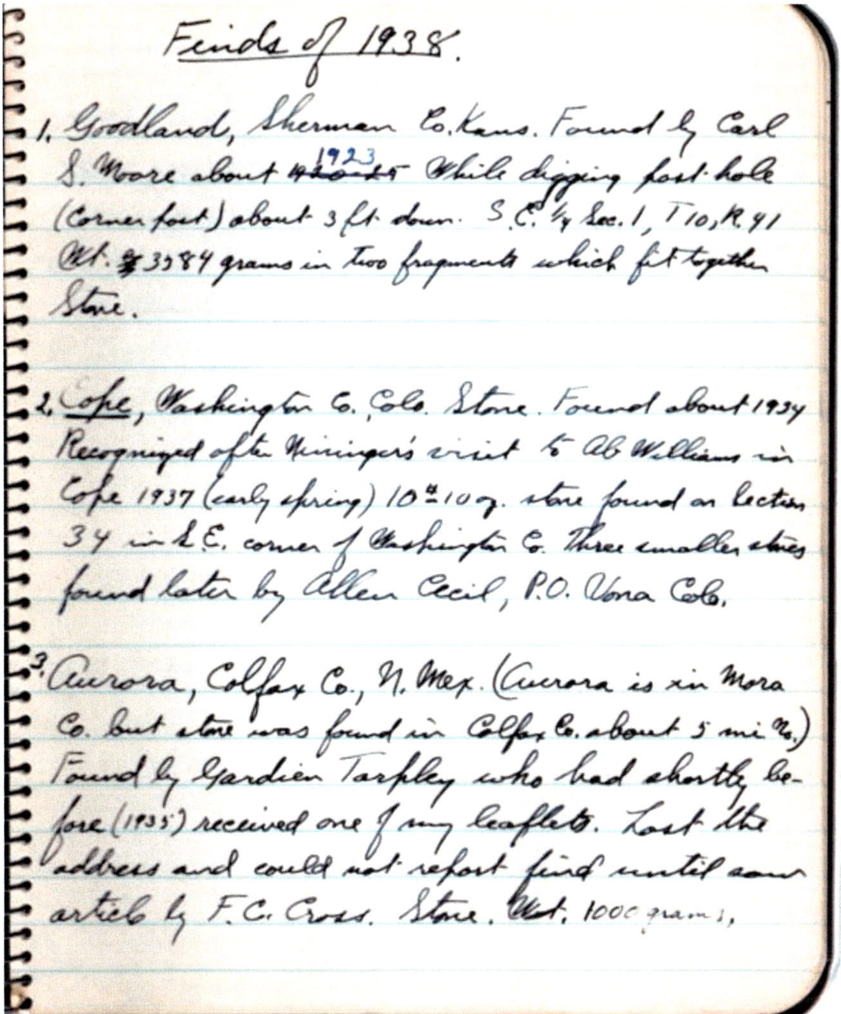

Finds of 1938.

1. Goodland, Sherman Co. Kans. Found by Carl S. Moore about 1923 While digging post hole (Corner post) about 3 ft. down. S.E. ¼ Sec. 1, T10, R.41 Wt. 3384 grams in two fragments which fit together Stone.

2. Cope, Washington Co., Colo. Stone. Found about 1934 Recognized after Nininger's visit to Al Williams in Cope 1937 (early spring) 10.10 g. stone found on Section 34 in S.E. corner of Washington Co. Three smaller stones found later by Allen Cecil, P.O. Vona Colo.

3. Aurora, Colfax Co., N. Mex. (Aurora is in Mora Co. but stone was found in Colfax Co. about 5 mi N.) Found by Gardien Tarpley who had shortly before (1935) received one of my leaflets. Lost the address and could not report find until saw article by F.C. Cross. Stone. Wt. 1000 grams.

Figure 2.32 Page from Nininger notebooks describing some of his 1938 meteorite collecting. Courtesy ASU Buseck Center for Meteorite Studies.

in Los Angeles, who then informed Dr. Frederick Leonard, Chairman of the Department of Astronomy at UCLA.

Alter informed Schmidt that because of the Antiquities Act, a permit would be required to remove the meteorite from the National Forest. Schmidt wrote to the U.S. Forest Service. The letter was passed on to the U.S. National Museum. A decision was made to approve the removal of the meteorite for transport to the National Museum. Schmidt was offered a $200 finder's fee. He was also advised that it would be advisable to have a Smithsonian scientist join the recovery team.

On April 3, 1939, Harvey and Addie Nininger went to California to give lectures about their meteorite collection. While in California he met with Leonard. As discussed in Chapter 1, Leonard and Nininger were close friends and shared a passion for meteorite study. Leonard told Nininger about the meteorite and his desire to keep the meteorite in California. Based on Leonard's suggestion, Nininger contacted Schmidt. Since Schmidt had made an arrangement with the Smithsonian, he refused to divulge the location of the meteorite to anyone outside of the Smithsonian.

In the early 1930s, Nininger had an arrangement with the Smithsonian to purchase up to $2,000 of his meteorites each year, however, this arrangement was terminated as the effects of the Great Depression deepened, although they occasionally purchased some meteorites from him. In fact, the Smithsonian paid Nininger $9,570 for 102 specimens from 81 meteorites between 1932 and 1939.[23]

When Nininger met with Schmidt, he apparently mentioned his association with the Smithsonian. Since Schmidt was expecting a Smithsonian scientist to accompany him in retrieval, he accepted Nininger and took him to the meteorite location. Leonard joined them on May 3. Upon reaching it, Nininger's excitement was evident: "Now we all joined in to fondle the prize, an almost unbelievable chunk of metal that seemed to link us, there in that wild corner of the earth, with worlds beyond."

The next day, Nininger wired the Smithsonian that he feared vandals would mutilate the meteorite and offered his services to remove it and ship it out for $70. The Smithsonian agreed but expressed surprise on receiving his telegram since they were unaware of his involvement.

Nininger arranged for several locals to rig up a four-ton log wagon, drawn by heavy draft horses. With block and tackle and other tools, bedding, provisions and cameras, they set out for the long hike. The workers had constructed a tripod of timbers from the surrounding forest. They tried to slide, haul and push the meteorite up a crude ramp to the wagon bed. Over two days, May 3 and 4, 1939, and after several breakdowns and repairs, the meteorite was finally loaded for the trip down the mountain (Figure 2.33) to the highway. It was then picked it up in a Mills College panel truck (Figure 2.34) and driven to the Chabot College Observatory. From the Chabot Observatory, it was taken to the Golden Gate International Exposition on Treasure Island in the San Francisco Bay.

Figure 2.33. Goose Lake meteorite loaded on to cart with Addie Nininger and Dr. Leonard. Courtesy of the American Meteorite Laboratory Photo Archives, Collections Research for Museums, Denver, CO.

Figure 2.34. Goose Lake meteorite viewed by the public. A Mills College van is shown with a banner announcing its arrival. Courtesy of the American Meteorite Laboratory Photo Archives, Collections Research for Museums, Denver, CO.

The Smithsonian had approved its temporary display at the Exposition.

As seen in Figure 2.35 at the Smithsonian, the meteorite is an octahedrite in an irregular mass that is deeply pitted. The dimensions are 46 in x 28 in x 20 in (1.22 m x 0.71 m x 0.51 m). When found, it rested "in the center of an almost circular, saucer-like depression or 'crater,' about five feet in diameter and one foot deep."[24] At the time, it was the largest known California meteorite. However, since that time the Old Woman meteorite, found in 1976 in the Old Woman Mountains, now holds the record of the largest California meteorite weighing in at 2.75 tons (2,750 kg).

During the time that the meteorite was on loan from the Smithsonian, and on exhibit at the San Francisco Exposition, considerable pressure from various California individuals and institutions was put on the Smithsonian to keep the meteorite in California. Despite this pressure, it was accessioned into the Smithsonian's National Collection of Meteorites and eventually shipped to the Smithsonian in 1941. The history of the Goose Lake meteorite affirmed the Antiquities Act and the disposition of meteorites found on Federal lands, and set the stage for later court rulings involving meteorites.

Figure 2.35. Goose Lake meteorite went on display in the meteorites hall. The hall was dismantled in the mid-late 1950s, as part of the Exhibits Modernization Program. It is still on display today. Courtesy Smithsonian Institution Archives, Record Unit 95, Box 44A, Folder: 5.

The War Years

The arrival of World War II in late 1939 until 1945 brought an end to Nininger's methods of meteorite hunting, since gasoline and tires were unavailable for anything other than the war effort. Rationing of food and supplies also limited his "spur-of-the-moment investigations." Extended travel was also curtailed. But by this time, Nininger's collection was well known and his reputation as a major collector was established. Fletcher Watson, of the American Museum of Natural History, wrote in his 1941 book *Between The Planets,* "H.H. Nininger, a modern pioneer in locating meteorites, has also been outstandingly successful . . . For several years Nininger was accounting for half of all discoveries in the world."[25]

Nininger was fifty-four years old when the war began. He was able to obtain full-time work for a time as a "salvage investigator" for the War Production Board. This involved looking for metals in old back country mine towns, along abandoned rail beds, and farm lots. From 1943 to 1946 he obtained employment exploring possible oil fields in New Mexico, and negotiating leases. These jobs gave him some opportunities to look around on his own during those trips.

For example, the **Johnstown**, Colorado, meteorite was a witnessed fall on July 6, 1924. Traveling through the area in 1939 or 1940, Nininger met Sam Dreith who was raking hay at the time of the fall. In one of his notebooks, Nininger recorded Dreith's experience:

> "Terrible commotion up in the sky; terrific bombardment; intermittent explosions; screaming, humming, popping and dull thud. Sounded like an airplane with missing motor. Intermittent. Horses tried to run; chickens cackled and ran to cover. Screamed followed by a thud. Looked around and saw puff of dust. Next day ran into it with rake. Stopped, saw it, then after dark went out and found it."

Nininger purchased the stone from him. It had been in his truck since the year he found it. The 1950 Nininger Collection catalog lists four Johnstown specimens.

In 1941, while in Bethune, Colorado, he saw three meteorites "mounted on a cement wall in a rock garden." He does not explain how he removed them from the cement wall, but there are three fragments of the **Bethune** meteorite in the 1950 collection catalog with a combined weight of 2.4 oz (0.68 kg). He also found a stony meteorite in a rock garden in **Abernathy**, Texas, weighing 6.4 lb (2.9 kg).

While driving through Wilmot, Kansas, in 1943, he stopped at a farm for some water. While getting a drink from a well, he noticed "a rusty-looking stone." Picking it up he realized that it was a weathered stony meteorite.

He paid $5 for the stone. The **Wilmot** meteorite was 4.4 lb (2 kg).

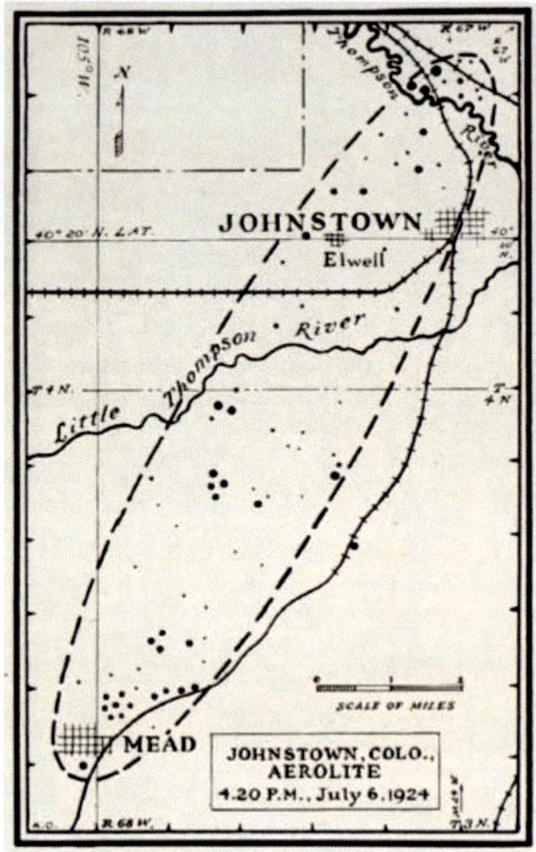

Figure 2.36. Map in the 1933 American Museum of Natural History Field Guide[7] of the Johnstown Shower.

In May, 1944, he was driving to Albuquerque when he stopped for a dinner break along the Rio Puerco. He recounts that he was strolling around the river bank with his "food in hand" indulging his fantasy of finding a meteorite just lying on the ground. After washing in the river, he was returning to his car when he notice that the area was dotted with dark pebbles. He spent the next hour scanning the pebbles. Giving up his search, he was heading back to his car when a particular dark pebble, sticking out of the sand caught his attention. He pried it out. A small break in the crust allowed an examination of its interior. He realized that it was a stony meteorite. The closest geological feature was Ladron Peak but it was found near the bridge (Puente) over the Rio Puerco. He therefore chose to name it the **Puente-Ladron** meteorite. It was a very small pebble, weighing only 0.3 oz (8.5 gm), that he believed may be just one of many, but he never returned to the area to look for others.

Despite the limitations placed on his travels and activities during the war, the 1950 collection catalog lists the following additions during that time:

1940 - Sixteen meteorites from Arizona, Colorado, Kansas, Oklahoma and Texas

1941 - Eight meteorites from Arkansas, Colorado, Nebraska, South Dakota and Texas

1943 - Two meteorites from Oklahoma and Nebraska
1944 - Two meteorites from Texas and Wyoming

1950 Catalog

As described in the earlier chapter, the end of the war was a time for a decision by Harvey and Addie on their future. His job with the oil company ended. He had decided to leave the Denver museum even before the war. Now was the time to follow through on that decision.

They had long thought about living near "the great Arizona meteorite crater," that they first visited on their 1925 Family Trip. In the summer of 1946, they moved their trailer to Flagstaff, Arizona, and decided to open a museum to show their collection. Between the new museum (Chapter 4) and their work at Meteor Crater (Chapter 3), their collecting activities were limited, yet they were able to add over 50 new meteorites through trade or purchase by the time they cataloged their collection in 1950.

The catalog[26] detailed 587 meteorites from thirty-nine states and forty foreign countries as of March 1950. To offer a comparison to the extent of this collection, the 1895 catalogue of the meteorite collection of

Alabama	6	Nebraska	27	Argentina	1	Japan	2
Alaska	1	Nevada	1	Australia	21	Java	1
Arizona	18	New Mexico	32	Belgium	1	Mexico	37
Arkansas	8	New York	4	Bolivia	2	Myanmar	1
California	7	North Carolina	12	Brazil	3	Nigeria	1
Colorado	39	North Dakota	2	Canada	8	Norway	1
Connecticut	1	Ohio	2	Chile	10	NZ	1
Florida	1	Oklahoma	7	Columbia	1	Peru	1
Georgia	7	Oregon	2	Czech.	5	Phillipines	1
Illinois	1	Pennsylvania	4	Egypt	1	Poland	2
Indiana	6	South Carolina	4	England	4	Portugal	1
Iowa	5	South Dakota	5	Estonia	1	Russia	7
Kansas	58	Tennessee	13	Finland	4	Saudi Arabia	1
Kentucky	13	Texas	57	France	17	Serbia	1
Louisiana	1	Utah	3	Germany	11	Spain	1
Maine	2	Virginia	5	Greenland	1	Somaliland	2
Michigan	5	Washington	2	Guatemala	1	South Africa	8
Minnesota	5	Wisconsin	7	India	13	Sweden	2
Missouri	14	Wyoming	9	Ireland	2	Ukraine	3
Montana	1			Italy	8	Yugoslavia	1

Figure 2.37. Table of meteorite locations from *The Nininger Collection of Meteorites*, 1950.

Chicago's Field Columbian Museum totalled 180.[27] The 1902 catalog of the collection of the United States National Museum totalled 348 falls.[28]

1954 Cottonwood Meteorite

In his notes,[29] Nininger describes how he was driving on a road to Cottonwood, Arizona, in 1954 when he noticed "a pile of blue-grey lava boulders and they definitely did not belong where I saw them. . . . At least a mile away was the nearest source for such boulders. Knowing that the aborigines sometimes covered meteorites with boulders, naturally the thought flashed through my mind that perhaps this pile concealed something of interest."

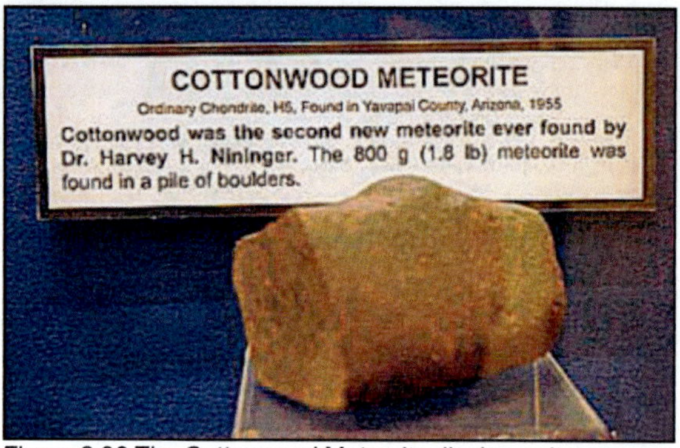

Figure 2.38 The Cottonwood Meteorite display as it appeared in the Sedona Meteorite Museum.

This reference to "aborigines," or Native Americans, remained in the folklore of this discovery but the author found that Nininger's typed notes clearly state that "I happened to glance down and at my very feet lay a small stony meteorite, the size of a large potato." This meteorite fell in the Coconino National Forest.

The display card shown in Figure 2.38 was for the **Cottonwood** meteorite in the Sedona Meteorite Museum (Chapter 4). The card states that this was only the second new meteorite found by Nininger. His excitement is obvious in his typed notes:

> *"I drove back to the museum, 14 miles, to report to Mrs. Nininger and the other members of our staff. For this exciting incident was too important to investigate further without letting my co-workers in on it. We should also confer on a plan for further investigation.*
>
> *After showing the find to all of our staff, Mrs. N. and I drove out again that she might see the lay out. For after all, the discovery of a completely new meteorite fall by a member of our own family is a rare and important event even after 32 years of devotion to this enterprise.*
>
> *Further search revealed more rock piles scattered over a half mile area. . . . A thorough search must be made. . . ."*

There are no accounts of whether they returned to the site later, but he did continue to search for information on the rock piles that he suggested may have been made by "aborigines."

> *"We continued to inquire and found an old man in the neighborhood who had the answer. He remembered when some 25 years ago the first road was being built through that area. He said the construction company had hauled several truck loads of waste rock that was too course for use in the roadbed and dumped them promiscuously without rhyme or reason. . . Well, this seemed to close the case of the rock piles having anything to do with the concealing of meteorites."*

This ordinary chondrite weighed 1.8 lb (0.8 kg) when inventoried. Since that time, five slices, totaling 0.3 lb (0.14 kg), were removed for exchanges, including the Chicago Field Museum and the American Museum of Natural History.

Fossil Springs Meteorite

The **Fossil Springs** Meteorite was discovered in the summer of 1945 by a worker who lived at Fossil Creek (Arizona) and patrolled the nearly 1.7 mile concrete flume that brought water to the Childs hydroelectric plant. According to information in the Buseck Center for Meteorite Studies files, this specimen is 11 lb (5 kg) and was purchased by Nininger. Little is known about how Nininger came to acquire the item, the price he paid,

Figure 2.39. Fossil Springs meteorite at ASU Buseck Center for Meteorite Studies.

or its exact location along the flume. A specimen of this size is not included in the 1950 collection catalog so it is assumed to have been acquired after the creation of the catalog.

Fossil Creek runs through the Coconino National Forest. There are numerous ancient dwellings on the ridges above the Fossil Creek flume, so it is regarded as being found "in the vicinity" of dwellings. It was identified by the ASU Buseck Center for Meteorite Studies as a Canyon Diablo meteorite fragment. The flume and power plan were removed in 2008 as part of the decommissioning of the plant. The area has now been designated as the Fossil Creek Wild and Scenic River Area.

CHAPTER 3

Meteor Crater

> *H.H. Nininger* and *Meteorites* go together, like word pairs such as *bread* and *butter*, at least for those who study these rare samples of cosmic debris.
>
> —- Fred L. Whipple

Meteor Crater was described in a 1933 publication on "Comets, Meteors, and Meteorites," by the American Museum of Natural History in New York[1] as follows:

> *The most remarkable occurrence of iron meteorite associated with a meteor crater is in Yavapai County, Arizona, near the intersection of the 111th meridian and 35th parallel. The meteorite, which is called Canyon Diablo, after a near-by erosion feature, consist of thousands of pieces of variable sizes scattered over an oval area about nine mile in diameter. A depression known as Meteor Crater lies in the center of the meteorite field. This crater is quite large being 4150 feet in diameter and 570 feet deep. It is surrounded by a parapet 150 feet in height composed of rock debris thrown out of the crater. The crater has been studied at various times from different standpoints and the present general consensus of opinion is that it was formed by the impact of a great meteor or comet with the earth some 50,000 year ago, and that immediately following the impact there was a tremendous explosion which not only scattered the meteorite and rock debris over the surrounding plain, but gouged out the crater pit and greatly disturbed the normal disposition of the thick limestone and sand stone bed in the margin of the crater. Borings have been made within and about the margins of the crater in an endeavor to locate a possible larger meteoric mass, but so far they have yielded only inconclusive results.*

This chapter explores the study of meteor crater, emphasizing the many years of research and discovery by Harvey Nininger.

The Wheeler Expedition

George Montague Wheeler, Army Corps of Engineers, was a young first lieutenant, having graduated from the United States Military Academy at West Point in 1866 at the age of 24. While stationed in San Francisco, Wheeler learned the art of military mapmaking. He led his first reconnaissance into Nevada in 1869 where he envisioned what would become known as the "Wheeler Survey."

In the spring of 1871, Wheeler assembled a group of scientists, soldiers, and surveyors (Figure 3.1) to participate in the first U.S. Army Engineer survey in the American West since the outbreak of the Civil War ten years earlier. The group's goal was to map the entirety of half of the United States, west of the 100th meridian. The members of the survey criss-crossed the West on foot, horseback, muleback, trains, and wagons. The Wheeler Survey lasted until 1879.

Alfred "Franklin" Banta served as a guide for the Arizona territorial government for almost six years when Wheeler's survey began. Banta was hired as the chief guide for the Wheeler survey. Banta discovered Meteor Crater during the Arizona portion of the survey. Wheeler named the crater "Franklin's Hole" after its discoverer.[2]

However, neither the "discovery" nor the name "Franklin's Hole" was recorded in the reports of Wheeler's survey, and neither the crater nor a "Franklin's Hole" was noted on maps prepared as a result of the survey's work in Arizona. This name does not appear on maps made over the next few decades.

Figure 3.1. Members of the Wheeler Survey c.1874. Courtesy of the National Portrait Gallery, Smithsonian Institution.

By Any Other Name

Meteor Crater has been known by several names. In addition to "Franklin's Hole," sometime in the 1880s the crater was known as "Coon Mountain" or "Coon Butte." These names were used by locals and in scientific publications well into the 20th century. It is, of course, neither a mountain nor a butte, and there is no record of racoons being in the area. However, the feature does give the appearance of a butte from a distance. Figure 3.2 is a drawing by Harvey Nininger of Meteor Crater as it is approached on foot.

Figure 3.2. Drawing of Meteor Crater by H. H. Nininger in his book "Arizona's Meteorite Crater: *Past, Present, Future*," 1956.

The origin of the name Coon butte or mountain is unknown; but in 1906, after making local inquiries, geologist H. L. Fairchild[3] attributed it to herders who used the low topographic relief of the crater's rim for a landmark for a nearby watering hole called "Coon Tank," which does not help in determining the origin of "Coon."

Interest in this geological feature began in March 1891 when a prospector sent a letter and sample to a mining company in Albuquerque, New Mexico, stating that he had found a vein of metallic iron near Canyon Diablo, Arizona. A sample also made its way to the land commissioner of the Atlantic and Pacific Railroad in Chicago, who sent it to A. E. Foote. Foote (1846–1895) was a Philadelphia physician, mineralogist, and mineral dealer. Foote reported that "A glance at the peculiar pitted appearance of the surface and the remarkable crystalline structure of the fractured portion convinced me that the fragment was part of a meteoric mass. . . As soon as possible, in June, I made a visit to the locality."[4]

Unfortunately, Foote added a bit to the naming confusion by identifying it from a U.S. Geological Survey map as "Sunset Mountain." Sunset Mountain is nine miles to the southeast of the crater.

Foote visited the site with five other men. About two miles from the floor of the crater, they found the two largest meteorite specimens at 201 lbs (91.2 kg) and 154 lbs (69.9 kg). As they explored the area they gathered 131 small masses ranging in weight from 0.06 oz (1.7 g) to almost 8 lbs (3.63 kg). At the base of the crater they collected a total of 200 lbs (90.7 kg) of oxidized and sulphuretted fragments that he concluded were "undoubtedly of meteoric origin." Yet despite this conclusion, he stated that "A careful search, however, failed to reveal any lava, obsidian or other volcanic products. I am therefore unable to explain the cause of this remarkable geological phenomenon."[5]

Following Foote's visit, local interest in the site increased and traders in the area began to collect and sell over a ton of meteoritic materials. The most notable trader was Fred W. Volz, proprietor of the Canyon Diablo Trading Post, so the specimens became known as the Canyon Diablo irons. Foote presented the results of this first scientific investigation of Coon Mountain and its "crater" to a meeting in Washington, D.C. on August 20, 1891. The first inclusion of the crater as "Coon Mt" in a U.S. Geological Survey topographical map[6] was found in the 1894 edition (Figure 3.3).

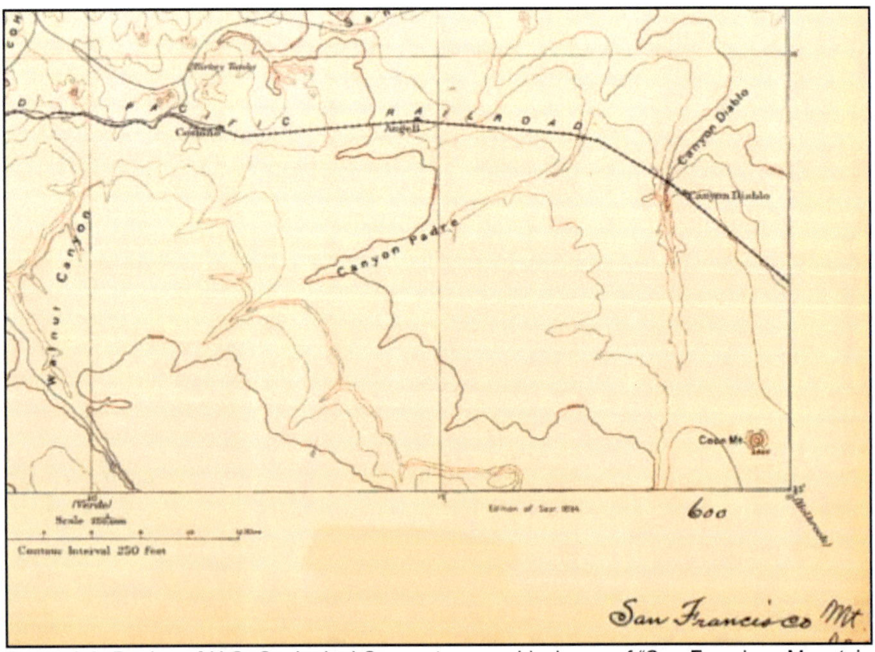

Figure 3.3. Portion of U.S. Geological Survey topographical map of "San Francisco Mountain Sheet," 1894 edition, showing "Coon Mt" in lower right corner. Courtesy U.S. Geological Survey.

The name "Meteor Crater" was first formally proposed by H. L. Fairchild, then secretary of the Geological Society of America, in his 1906 paper.[7] Fairchild had visited the crater earlier that year and had become convinced of its impact origin. Fairchild argued that it was not a butte and its height did not bring it to the level of a mountain. He thought "crater" was more accurate. Likewise, he proposed "meteor," both because of the numerous siderites associated with the crater, and because the nearest post office was named "Meteor." But in 1908, George P. Merrill, head curator of geology for the U.S. National Museum, declared that crater was "a singularly inappropriate use of terms."[8]

Despite Merrill's objection to the term, the *Guidebook to the Western United States*[9] produced by the U.S. Geological Survey in 1915 included mention of the crater: "about 10 miles south of Sunshine is Crater Mound, long known as Coon Butte and for a while as Meteorite Mountain, perhaps the most mysterious geologic feature in the West."

Arizona Highways magazine contributed a little to the naming confusion in a September 1938 issue with an article entitled "Meteor Mountain"[10] by adding another name to the site. "Of the many natural wonders located in northern Arizona including the Grand Canyon, the Petrified Forest, the Painted Desert, and Meteor Mountain, or as it is often called **Meteor Mine**, none is of more interest to scientists than Meteor Mountain" (Emphasis added).

The name Crater Mound was used widely in most publications until 1946 when the name, Meteor Crater, was officially adopted by the U.S. Interior Department.[11] The Meteoritical Society refers to the crater as Barringer Meteorite Crater[12] because, as will be discussed shortly, Daniel M. Barringer was one of the first people to suggest that it was produced by a meteorite impact. They also point out that, technically, it should be called "Meteorite" Crater since a meteor becomes a "meteorite" once it survives earths atmosphere and lands on the ground.

Under the provisions of the Historic Sites Act, a National Landmark is defined as a site that possesses exceptional values in illustrating the natural history of the United States. "Barringer Meteor Crater" was designated a National Landmark in 1967[13] by the U.S. Department of the Interior, National Park Service, with a plaque being presented in 1968. There were only 602 designated National Natural Landmark sites in the United States and territories as of 2022.

The science of meteoritics is born

The subject of meteorites and possible craters created by their fall was controversial for centuries. "Mysterious heaven stones," "thunderstones," "firestones," "shooting stars" are some of the terms used to describe meteorites over the centuries.

A German physicist, Ernst Chladni, was the first to suggest that meteorites came from outer space. Chladni became interested in meteoritics following a conversation with a witness to a November 1792 fireball. Chladni researched reports of similar witnessed falls across Europe and North America. Based on the similarities of these sightings, he concluded that the phenomena must be taken seriously.

When Chladni published a book[14] on his findings in 1794, he was roundly criticized. However, attitudes began to change after a witnessed fall of over 3,000 meteorite fragments in the French town of L'Aigle on April 26, 1803. Upon hearing of this event the French Academy of Sciences sent the scientist Jean-Baptiste Biot to investigate. After extensive work in the field he reported two kinds of evidence pointing to an extraterrestrial origin for the stones:

1. Physical evidence: the sudden appearance of many identical stones similar to other stones fallen from the sky in other places
2. Moral evidence: many witnesses observed a "rain of stones thrown by the meteor."

Biot's paper[15] describing how these stones must be of extraterrestrial origin effectively gave birth to the science of meteoritics. The L'Aigle event was a milestone in the understanding of meteorites and their origins.

While it was acknowledged after 1803 that meteorites came from outer space, the idea of a meteor impact crater on earth was still not acknowledged. At the same time, there was significant debate on the origins of the craters on the moon. Most astronomers attributed lunar craters to volcanic activity, although no lava flows were detected. In 1873, Richard Proctor published "*The Moon: Her Motions, Aspect, Scenery, and Physical Condition*"[16] in which he proposed that the craters were the result of "meteoric" impacts. However, recognizing the criticism that this theory could generate, he softened his position by recognizing that it might be "too wild and fanciful to suggest." In fact, when his book was updated[17] in 1878, this theory was removed. The volcanic eruption theory continued to be the prominent belief for decades.

Figure 3.4. Plate XX. "Portion of the moon's surface from a model by Nasmyth in Richard Proctor's book "The Moon."

Grove Karl Gilbert

Among those attending Foote's 1891 presentation, aforementioned, was Grove K. Gilbert. Gilbert was the chief geologist of the U.S. Geological Survey at the time. He is considered one of the most distinguished American geologists. Gilbert indicated that "the thought of examining the scar produced on the earth by the collision of a star was so attractive that I desired to visit the crater." However, he initially sent a colleague, Willard D. Johnson, who spent several days there making a sketch map and describing the various features. Johnson concluded that a body of steam was produced by volcanic heat, some hundreds or thousands of feet below. The steam caused an explosion to produce the crater. Concerning the surrounding iron fragments, he concluded that these were independent of the explosion, and the association with the crater was "accidental."

Figure 3.5. This portrait of Grove Karl Gilbert (1843-1918) was done in the 1920's by Abner Lowe. Courtesy U.S. Geological Survey.

Johnson's report increased Gilbert's desire to visit the site. He arrived at the crater in November 1891 with Marcus Baker, of the U. S. Geological Survey. He noted that "the crater is locally known as Coon Mountain, or Coon Butte. The iron is known to literature as the Canyon Diablo fall of meteorites."

On Saturday, November 7, Gilbert's field notes[18] indicates that he and Baker set up a plane table on the edge of the crater. They spent two weeks conducting topographic and magnetic surveys, as well as studies of the local details (Figure 3.6). As he approached the feature he noted that:

> "the plain is interrupted by a bowl-shaped or saucer-shaped hollow, a few thousand feet broad and a few hundred feet deep; and about this hollow is an approximately circular rim rising one or two hundred feet above the surface of the plain. In other words, there is a crater; but the crater differs from the ordinary volcanic

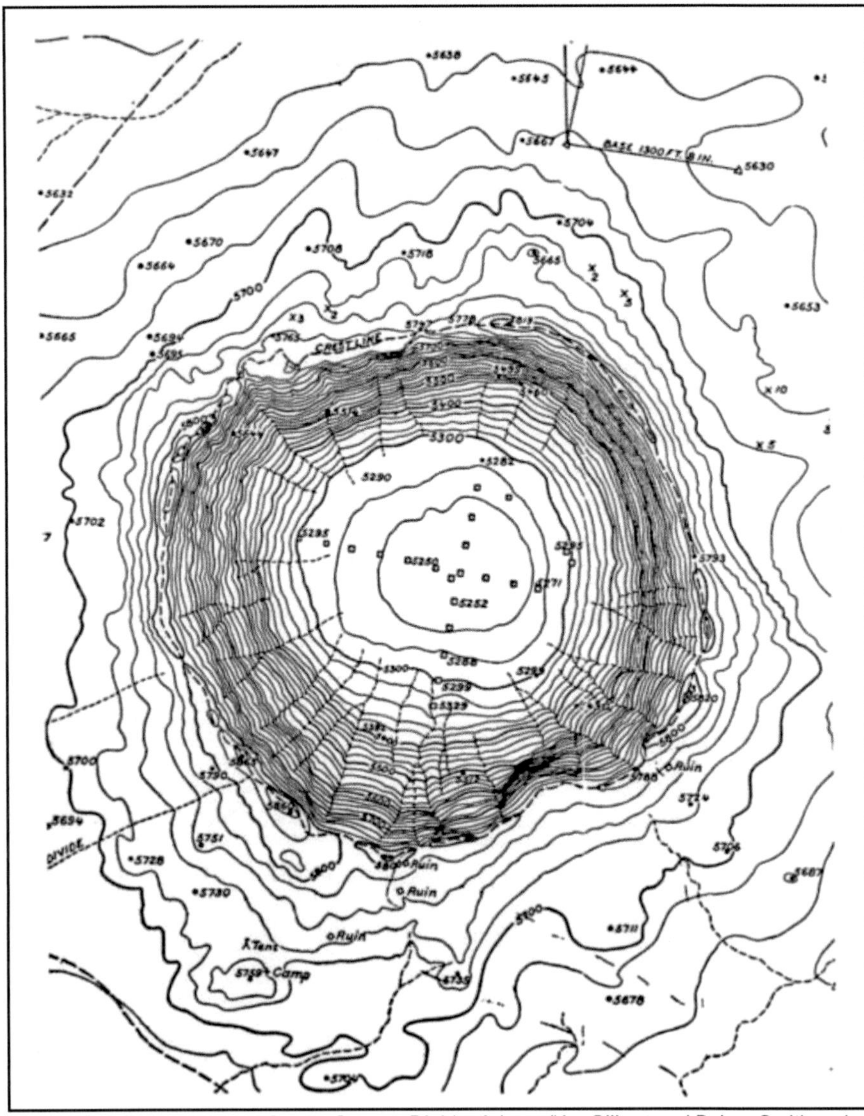

Figure 3.6. "Contour Map of Crater, Canyon Diablo, Arizona" by Gilbert and Baker. Smithsonian Miscellaneous Collections. Copy provided by U.S. Geological Survey, Flagstaff.

structure of that name in that it contains no volcanic rock. . . On the slopes of this crater and on the plain round about many pieces of iron have been found. . . ."[19]

They performed several tests to determine the cause for the crater. He calculated the amount of material that would have been ejected to create the crater and compared that to the surrounding landscape. If there were a large meteorite in the crater, there would be "extra" material in the surrounding landscape. He found no such extra material.

He also tested the iron meteorites. A compass needle reacted to irons outside the crater, but they found no magnetic compass effects on the floor of the crater. He observed that "the theory of a great iron meteor is negatived by the magnetic results, unless we may suppose either that the meteor was quite small as compared to the diameter of the crater, or that it penetrated to a very great depth." The two experiments worked against the theory of a colliding meteor. Gilbert concluded that he agreed with Johnson that the crater was caused by underground volcanic magna that had heated a pocket of water, resulting in a steam explosion.

Daniel Moreau Barringer

In a 1951 *Science Monthly*[20] article, Nininger recognized the contribution of Daniel M. Barringer to the debate on the origin of Meteor Crater.

Figure 3.7. Daniel Moreau Barringer

"It is to D. M. Barringer and his associates that must go the credit for first recognizing and more or less establishing the meteoritic origin of the crater, and to them we are also indebted for a great body of facts regarding the structure and general characteristics of the phenomenon."

Daniel Moreau Barringer was born in Raleigh, North Carolina. After his father's death, he was taken in by his uncle in Philadelphia. Barringer was admitted to Princeton University at the age of 15 into the class of 1879 which included Cyrus McCormick and Woodrow Wilson. He graduated when he was 19 years old. His uncle was a lawyer so he then entered the Law School of the University of Pennsylvania, and graduated at the age of 22.

He apparently was less interested in the field of law than in big game hunting in the West and Canada. With his friends, including Theodore Roosevelt, he helped found the Boone and Crockett Club.[21] The organization continues to this day with its mission of promoting the conservation and management of wildlife, especially big game, and its habitat, and to preserve and encourage hunting and to maintain the highest ethical standards of sportsmanship in North America.

His interest turned to geology so he enrolled at the University of Virginia where he studied metallurgy. His legal and geological knowledge were combined in a book, *The Law of Mines and Mining in the United States* (Barringer and Adams 1897).

Barringer came to Arizona when he met and married a woman from Arizona. With some partners, they found a silver mine near Pearce, Arizona. Perhaps it was fate that this move and mine brought meteor crater to his attention. As described by his son Brandon,[22] his father personally operated the mine in 1902 when he met Samual Holsinger. Holsinger was a surveyor and geologist with the U.S. Forest Service. Holsinger came down to inspect the Pearce mine, to be sure that the mine timbers were not being cut from the National Forest. "Over the campfire that night, Holsinger told Father a story of a large crater in northern Arizona with meteorites lying all over the rim." Holsinger was relating stories that he had heard even though he had not been to site himself.

Barringer later reported[23] that "I naturally was very incredulous of the theory which, Mr. Holsinger informed me, was held by some of the people living in the neighborhood of Canyon Diablo, namely, that this great hole in the earth's surface had been produced by the impact of an iron body falling out of space. . . ." But the subject continued to interest him. In January 1903, he met his friend, Benjamin C. Tilghman at the Boone and Crockett Club in Washington, D.C. Tilghman was a physicist, chemist and mathematician. They were both intrigued by the prospect. Barringer wrote Holsinger asking him to check the story he had related and to send them "slivers" of the meteorites to examine. On February 17, having the answer and samples from Holsinger, Barringer wrote that

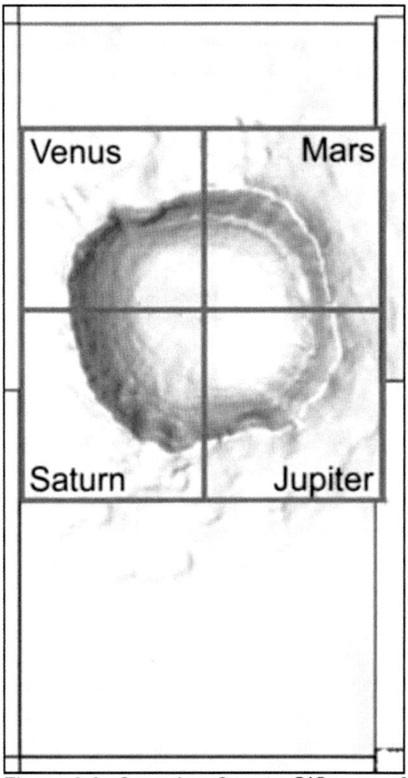

Figure 3.8. Coconino County GIS map of Meteor Crater's current 1,280 acres. The four mining claims are show with their approximate boundaries.

they had decided to acquire the property.

Barringer was well known in the Arizona mining community. He, therefore, did not want to be publicly linked to the site until all claims had been secured for fear of claim jumpers. The area around the crater was unsurveyed public lands. Four placer mining claims, totalling 640 acres, were filed in the name of Holsinger and three of Barringer's brothers-in-law who also had initial interests in the enterprise. The mining claims were named Venus, Mars, Jupiter and Saturn. A placer claim grants to the discoverer of valuable minerals, contained in loose material, such as sand or gravel, the right to mine on public land.

Holsinger was hired away from the Forest Service to file the claims and supervise the work needed on the property to fulfill the requirements for patenting the mining claims. A land patent is a legal document issued by the federal government that grants an individual title to land that is held by the government. These patents served as the early forms of title, and they allowed people to assert their ownership against all others. A trail was constructed to the crater and shallow shafts were being sunk into the crater's floor to meet the patent requirements.

Back in Philadelphia, Barringer chartered the Standard Iron Company to which the placer claims were assigned. He then began the process of obtaining a patent for the land, which was granted on December 24, 1903, and signed by his hunting friend, President Theodore Roosevelt.

Standard Iron Company

The first exploration of the crater was a central shaft, at the juncture of the four placer claims. In September 1903, Holsinger was directed to sink shafts and run tunnels in the ejecta to "establish ... whether fragments of the meteor are admixed with sandstone and limestone fragments from the present surface down."[24]

Borings continued with two holes in 1904 and three in 1905. Despite not yet finding the meteorite mass, Tilghman and Barringer collected evidence in favor of the theory that "this great hole in the earth's surface" was caused by a meteorite to refute Gilbert's steam explosion theory. There was a concern that they needed to get ahead of any possible negative publicity that could jeopardize future fundraising for the project. They each presented papers of their findings on March 1, 1906.

Barringer's paper[25] presented several points that argued against the steam explosion or volcanic eruption theories:

① There are no eruptive rocks of any sort in the area to suggest volcanic origin of the crater.
② There is not a single point of similarity, excepting perhaps that of the round shape of the interior basin, to a volcanic eruption.
③ The many shattered and cracked exposed limestone fragments are showing that they have been subjected to a concussion from a great blow.
④ "Meteoritic material" has been found on the surface, including several large specimens, one weighing over 200 pounds and others over 100 pounds, having been found nearly a mile west of the crater, and many small ones distributed around it.

Tilghman's paper[26] presented several "facts" to support the theory that the formation of the crater was due to the impact of a meteor "of enormous and hitherto unprecedented size." Among the "facts" presented was that the surface of the rim and surrounding plain is a large quantity of meteoric material that is symmetrically distributed in a line passing through the center of this hole. He also pointed out that in and around the crater is an enormous quantity of pulverizer rock which can be produced by a "violent blow," but could not have been produced by natural erosion.

While Barringer and Tilghman developed and presented their papers, work at the crater continued with three shafts sunk in 1906. In 1907, sixteen more shafts were dug.

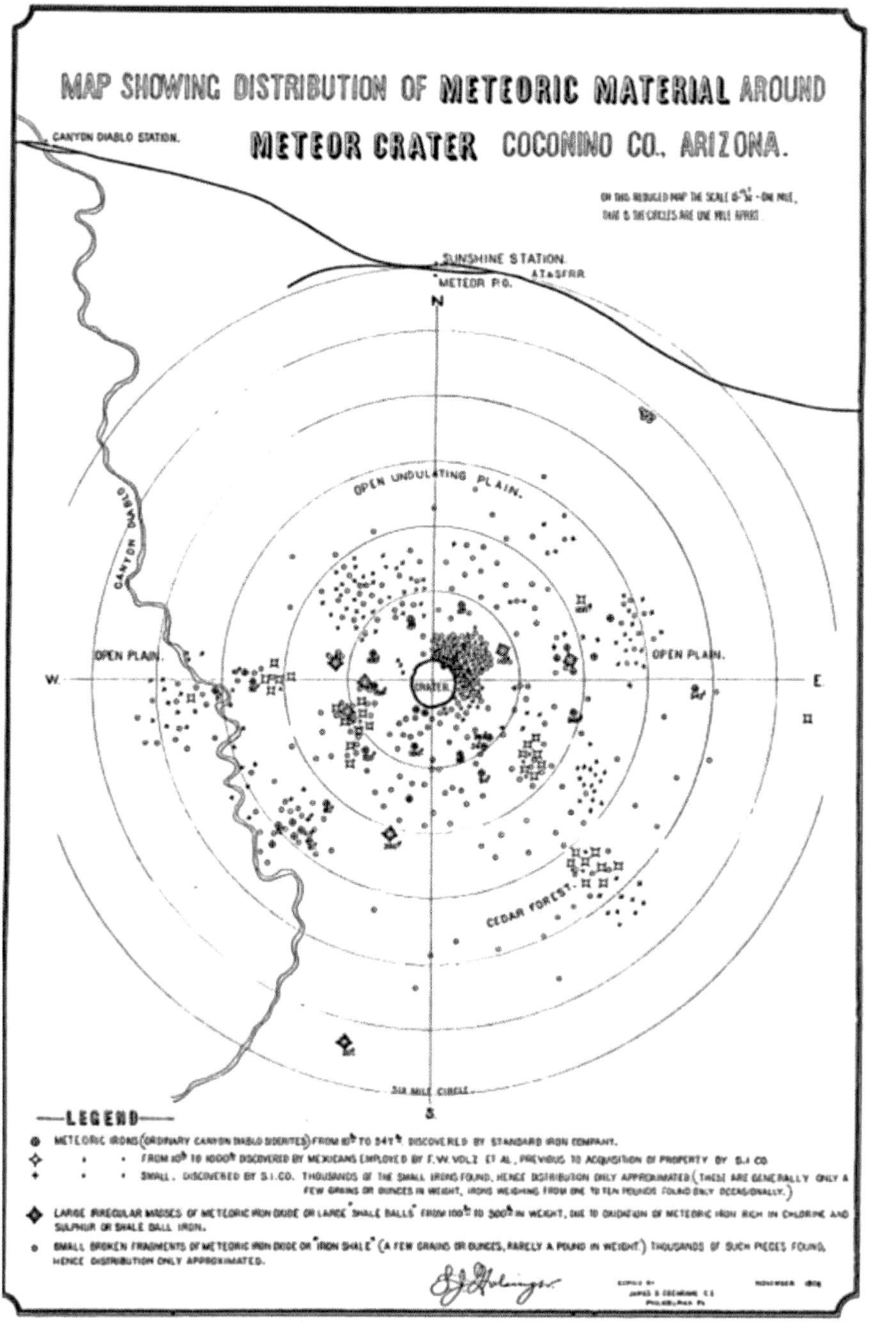

Figure 3.9. S. J. Holsinger's November 1906 map, prepared for the Standard Iron Company, showing the distribution of meteoric materials around Meteor Crater.

One person who obviously read the Barringer and Tilghman reports was George P. Merrill who, as mentioned earlier, had indicated that he thought calling the site a "crater" was inappropriate. However, his 1908 paper was entitled "The meteor crater of Canyon Diablo, Arizona; its history, origin and associated irons."[27] The paper reviewed all the studies that had occurred to that date, but he noted: "Though a matter of frequent discussion among those more or less conversant with the facts, nothing of value relating to the subject appeared until 1905, when Messrs. D. M. Barringer and B. C. Tilghman presented their results before the Philadelphia Academy of Science." He went on to suggest that "even the plausible suggestion that such might be due to the impact of a stellar body is of so unusual a nature as to warrant the fullest investigation." He concluded that "The work thus far done does not, therefore, disprove the presence of a large quantity of fragmental iron, although tending to show that no large mass lies there buried." Barringer was not totally happy with the report but he used Merrill's descriptions of the possibility of a massive iron meteorite as "plausible" and "does not, therefore, disprove" as incentives to continue their work.

Work did continue at the crater with four more shafts drilled by July 1908. By the end of that year, a total of twenty-eight shafts had been drilled over the years. All the shafts were dug in the central area of the crater. None of these attempts resulted in finding the massive iron meteorite that Barringer was sure was there.

The cost of the shafts was being borne primarily by Barringer and Tilghman. Tilghman pulled out of the project in late 1908 forcing the

Figure 3.10. Boiler and shaft house at location of the earliest drill holes. Courtesy Harry Locke Collection, Old Trails Museum/Winslow Historical Society.

Figure 3.11. Derrick inside crater where first exploration holes were drilled in the crater. Courtesy Harry Locke Collection, Old Trails Museum/Winslow Historical Society.

suspension of drilling for several years. During the suspension of operations, Tilghman passed away in 1910. Holsinger, who enticed Barringer with the meteorite story, passed away in 1911 at the crater. Barringer spent the next several years trying to raise funds for further exploration of the crater.

In 1918, Barringer was able to convince the U.S. Smelting, Refining and Mining Company to provide funding for additional drillings. Several additional shafts were drilled but no meteorite mass was found, instead, many of the shafts either struck water or quicksand. The drill in one shaft, at 1,367 ft (416.6 m), got stuck, which was interpreted as hitting a large meteorite. Despite this possibility, by the end of 1922, U.S. Smelting pulled out of the project. Finding additional investors proved difficult.

Barringer's son, D. Moreau "Reau" Barringer, Jr., who earlier presented a paper on the Odessa Crater (discussed in Chapter 2), wrote a series of articles for *Scientific American*[28] that appeared in the July, August and September 1927 issues. The three-part article was entitled "The Most Fascinating Spot on Earth." The article described the work that had been done at the crater and how explorations to find the main meteorite mass continued. Many photographs and a map made the subject of the article more approachable to a non-scientific audience.

These articles are credited with successfully acquiring additional investors. Not only did they appear in a respected publication, but the editor chose to become involved in the debate with the following inserted statement:

Science Backs Meteor Crater

Because certain people, reluctant to believe the unprecedented, regard as sensational the theory that Meteor Crater was formed by the impact of a giant meteor which struck the earth, we have obtained the following definite statements from two well-known scientists:

"I am perfectly willing to make a strong affirmative statement in support of Mr. Barringer's article," writes Dr. W. F. Magie of the Palmer Physical Laboratory, Princeton University, "but there ought to be no need for it. There is no reasonable doubt that the Crater was formed by the fall of a meteor and that this meteor is buried in it."

Dr. Elihu Thomson, Director of the Thomson Laboratory of the General Electric Company, writes, "I am very willing to be quoted as follows: 'There can be no question of the Crater being made by masses of meteoric iron, and that an enormous mass of such iron remains buried under the south wall of the Crater.'"

The Editor.

Figure 3.12. Insert on the first page of Barringer's *Scientific American* article[28]

These investors made it possible to resume drilling in 1928. However, in 1929, with no positive results, and the beginning of the Great Depression, investors became nervous. They hired Forest Ray Moulton, a professor of astronomy at the University of Chicago, to conduct further studies. Moulton filed two reports, both of which were devastating to Barringer. Using mathematical models, Moulton concluded that the meteorite was much smaller than Barringer believed, and that most of it fragmented and exploded away rather than leaving a large deposit within the crater. "It came as a bombshell late in August, and, despite all that father could do, killed the enterprise."[29] Dejected, Barringer went home for Thanksgiving and passed away of a heart attack on November 30, 1929.

At the time of his father's death, Brandon Barringer estimated that "well upwards of $600,000, exclusive of interest, had been spent on the exploration by the time of father's death, including over $120,000 of his personal funds."[30] Adjusted for inflation, $600,000 and $120,000 in 1929 is equal to $9,782,526 and $1,956,505, respectively in 2022.[31]

No further commercial explorations of the crater were undertaken after Barringer's death. However, debate on its origin continued for several decades, which drew Nininger's attention to the question.

The Crater and Nininger

A short handwritten note, in the files given to the Sedona Historical Society, by Nininger, is entitled "The Arizona Crater and Me."[32] The short, undated, page-and-a-half note provides a very brief overview of his activities at the crater. It begins:

> "I don't remember just when I first heard or read about Arizona's Meteorite Crater, but I first saw it in the spring of 1926 when we were winding up our 1925-26 Natural History Trek.* This was the school-year trip that the Nininger family made in our newly constructed house on wheels, starting in Aug. 1925 and ending in June 1926. We parked our house-car on the northwest rim, intending to stay a few days; but little Margaret developed a bad throat infection during the night and so we moved into Winslow next day in order to get her to a doctor."

In his book *Arizona's Meteorite Crater*, Nininger mentions that he spoke to Daniel Barringer "in 1928, at which time he expressed himself as absolutely certain that the main mass of the meteorite was resting under the southern rim of the crater."[33] In the note mentioned earlier, Nininger goes on to describe that after "we moved to Denver in 1930 I made several visits to the crater."

These early years (Chapter 2) showed Nininger's interest in meteoritic craters. The "buffalo wallow" at Haviland had fascinated him from his first visit in 1929. He was further intrigued by the Odessa crater during his 1932 visit, resulting in the 1933 article that declared it was formed by meteoritic impact. His return visit to Haviland in 1934, to excavate the site, was the first excavation to prove the existence of a meteoritic impact crater. It was therefore only natural that he would focus a good deal of his attention on Meteor Crater.

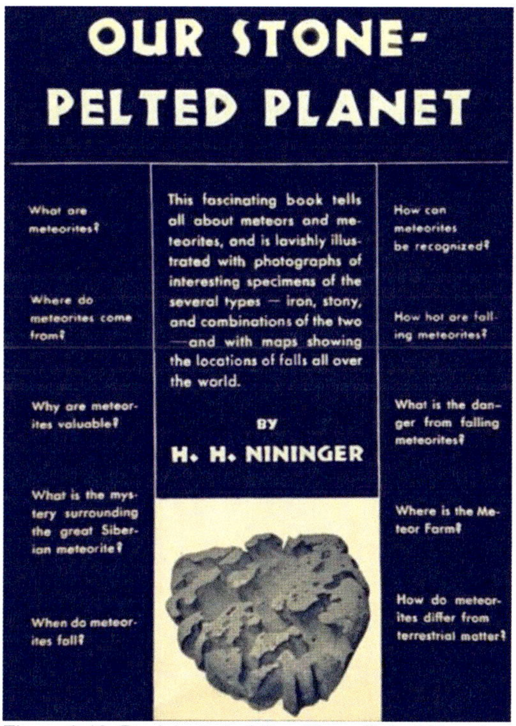

Figure 3.13. Book jacket of *Our Stone-Pelted Planet*.

* Nininger's recollection was in error in that the 1925-26 trip was referred to as the "House-Car Trip." The "Natural History Trek" was in 1927-28 with his students.

Figure 3.14. Pictured left to right, Mrs. Fales, Herbert Fales, photographer Otto Roach and Nininger in front of the plane in 1934 for the flyovers. All together they logged about 2,000 miles over four states. Note the large camera in the foreground. Courtesy Sedona Historical Society.

Figure 3.15. Photograph of Meteor Crater taken by Nininger during one of the flyovers. Courtesy of the American Meteorite Laboratory Photo Archives, Collections Research for Museums, Denver, CO.

After the Niningers settled in Denver, and during his study of impact craters, he began to write his first book, *Our Stone-Pelted Planet: A Book About Meteors and Meteorites*. It was published in 1933 when he was Curator of Meteorites in the Colorado Museum of Natural History and the Director of the Nininger Laboratory. He stated that the goal of the book was to attempt "to deal with the subject in a way to interest the reader without sacrificing accuracy."[34]

An April 1933 review by the *New York Times* said "he opens before their amazed eyes vistas of knowledge that appeal to the imagination like a Jules Verne novel. . . The lay reader desirous of information on this subject will find the book just what he wants and most interestingly written."[35] The Canyon Diablo meteorite is referenced six times in the book. In the chapter on the composition of meteorites, Nininger states that "It is the platinum content of the Canyon Diablo meteorites which has been largely responsible for the persistent attempts to reach the main mass of this great fall . . . such a content would, it was thought by some, render the mining of the large mass a profitable industry."

Shortly after the publication of this book, Nininger reports that he received a phone call from Herbert B. Fales, the Vice-President of the International Nickel Company. Fales had told Lewin Barringer, one of Daniel Barringer's sons, about some "elliptical scars" that he observed over the Carolinas. Lewin was a renowned aviator at the time and probably heard his father's Arizona crater stories. Lewin suggested that Fales contact Nininger. Fales offered Nininger the opportunity to "fly over the plains in search of some meteorite craters." Fales came out that summer with his plane and took Nininger on a 3-day flight looking for impact craters over Colorado, New Mexico and Texas. Figures 3.14 shows Nininger in front of the plane with other members of this first flyover.

Fales returned two years later to do a flyover of southern Colorado, northern New Mexico, and Meteor Crater. Figure 3.15 is the image of Meteor Crater taken in 1936 during that flyover. They circled the crater several times and landed in Winslow to visit the Crater. Fales became an amateur meteorite collector and purchased several specimens from Nininger over the years. It is interesting to note that the first known photograph of Meteor Crater from the air was by Charles and Anne Lindbergh during their famous 1929 flyover of archaeological and geological features.

As discussed in Chapter 2, Nininger spent the years from his 1930 move to Denver, writing, traveling, lecturing and collecting meteorites. He made several trips to the crater to collect specimens and to study the site. Nininger is credited with several very important discoveries during his studies. The balance of this chapter highlights some of the more significant discoveries.

1936 Canyon Diablo 2 and 3[36]

In 1936, while etching several small Canyon Diablo irons that had been purchased from Harry Locke of Winslow (who will be discussed further in Chapter 4), two irons were found to respond in a different manner to the etching process that uses nitric acid to reveal the crystalline structure of the meteorite. Nininger recognized these irons as being different from any of the previously known types of Canyon Diablo irons. Its content of nickel was about 8.4%, in contrast with the 5% to 7% of nickel which had been reported in Canyon Diablo irons. Nininger suggested that this new type of nickel-iron meteorite from the region of Meteorite Crater be called Canyon Diablo No. 2. Later, Nininger studied and catalogued another new type that he named, Canyon Diablo No. 3.

Nininger suggested that, since these two new types were found on the crater rim, that either there were multiple falls in the same area, or that these two new types became attached to the main mass during its orbit around the sun. A later study of the irons suggested that there "is no doubt that Canyon Diablo 2 and 3 are truly distinct from Canyon Diablo 1."[37] The possibility of multiple falls in the same area is rather remote. However, it is conceivable that they were formed by yet unknown processes in play as the meteorite traveled eons through space. Nininger's discoveries opened another door for future studies.

1939 Magnetic Rake

In 1939 Nininger believed that the factual data on the crater was inadequate and that a thorough study of the distribution of meteoritic materials around the crater would help to interpret what happened to create the crater. He spent the next sixteen years around the crater collecting and analyzing data.

Nininger applied for a grant from the American Philosophical Society in

1939 to study the distribution pattern of the meteorite fragments. A matching grant of $500 was awarded to the American Meteorite Laboratory to conduct this field work.[38] The grant required that an equal amount be raised from other sources. Dean Gillespie, for whom Nininger worked as a truck driver, came through and matched the grant amount.

To collect samples around the crater, Nininger constructed a "magnetic rake" (Figure 3.16) attached to the family car. Nininger, with son Bob, and employee Irving Hoaag, proceeded to the crater to search for very

Figure 3.16 Photo of Nininger's magnetic rake attached to family car. Courtesy of the American Meteorite Laboratory Photo Archives, Collections Research for Museums, Denver, CO.

Figure 3.17. Meteorite filings picked up by a magnetic rake. Courtesy of the American Meteorite Laboratory Photo Archives, Collections Research for Museums, Denver, CO.

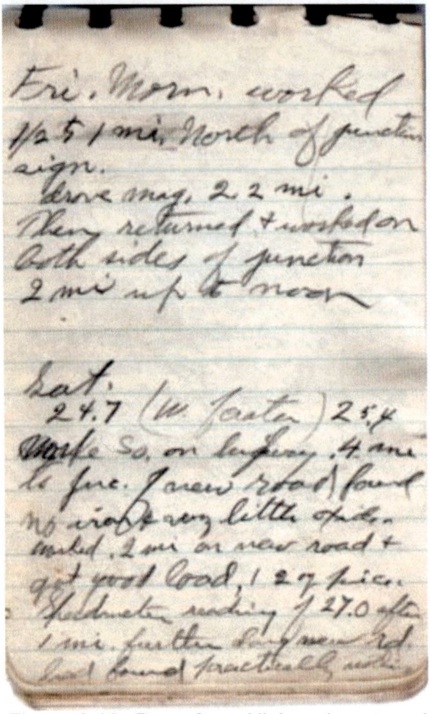

Figure 3.18. Page from Nininger's notepad labeled "Magnetic Rake." Courtesy ASU Buseck Center for Meteorite Studies.

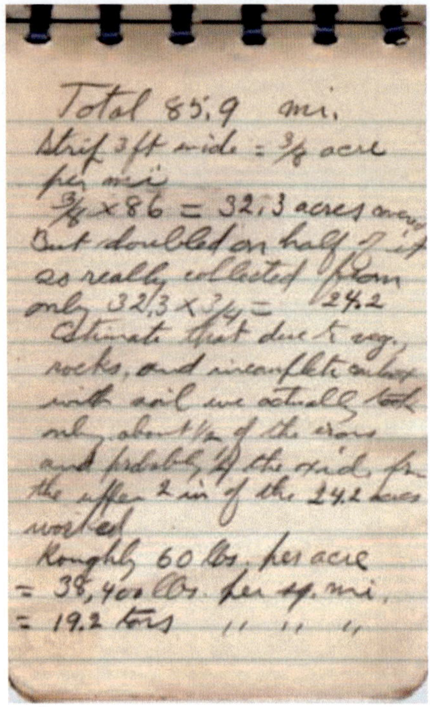

Figure 3.19. Summary page from Nininger's notepad labeled "Magnetic Rake." Courtesy ASU Buseck Center for Meteorite Studies.

small fragments as their objective. They began with a search of selected areas on various sides of the crater within a radial distance of 103 miles, using the magnetic rake. The rake was designed to gather small masses of an ounce or less, in the upper three-quarters-inch of soil. However, it did collect some larger specimens.

The area of their search was rather restrictive due to the uneven terrain, boulders and vegetation. They did find some workable terrain on each side of the crater but could not work an equal distribution on all sides. They found nothing farther than two and one-half miles from the rim of the crater. However, they found a strong tendency for the specimens to radiate from the rim as though they were thrown from the crater.

They worked the area for two weeks. Figure 3.18 is a page from the Nininger notebook labeled "Magnetic Rake." It details work performed on a Friday and Saturday. Figure 3.19 is the summary sheet at the end of the notebook. He estimates that they covered 85.9 miles. It describes Nininger's frustration with the survey in that "due to veg., rocks, and incomplete contact with soil we actually took only about 1/2 of the irons and probably 1/4 the oxide from the upper 2 in of the 24.2 acres worked." Due to these limitations, Nininger concluded that the survey was not extensive enough to warrant any conclusion other than that further surveys were required. He was, however, able to report isolating about 12,000 small meteorites with a total weight of just over 42 lbs (19 kg). He kept detailed records of these finds which revealed a radial distribution, indicating to him that they had been thrown from the crater.

1942 *A Comet Strikes the Earth*[39]

A Comet Strikes Earth, How to Recognize Meteorites, Visitors From Space was first published in 1942 by Nininger's American Meteorite Laboratory in Denver. The first edition had a blue cover of 40 pages with 7 black and white photos, and a centerfold map of the area around Meteor Crater. It described the Canyon Diablo meteorite and crater, and included a section on how to recognize a meteorite. There is even one of Nininger's poems entitled "A Message From Space." The "pamphlet" measures 6 x 4.25 in (about 11.0 x 15.0 cm). A feature of the pamphlet was an attached fragment that was identified by Nininger as "a genuine meteorite which was collected by the author within two miles or less' of the Crater rim.

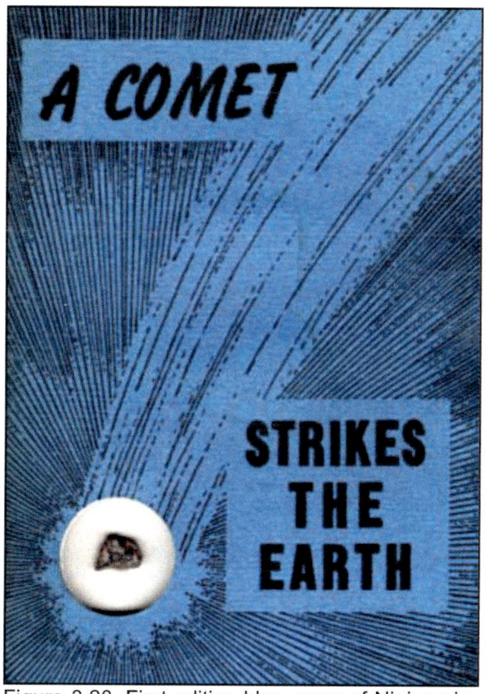

Figure 3.20. First edition blue cover of Nininger's 1942 *A Comet strikes the Earth* pamphlet.

The inclusion of a specimen created some controversy with the Meteoritical Society. The 1946 revision (with a gold cover), published by the American Meteorite Museum, Winslow, was expanded to 57 pages with 20 illustrations, plus a map of the area showing the crater.

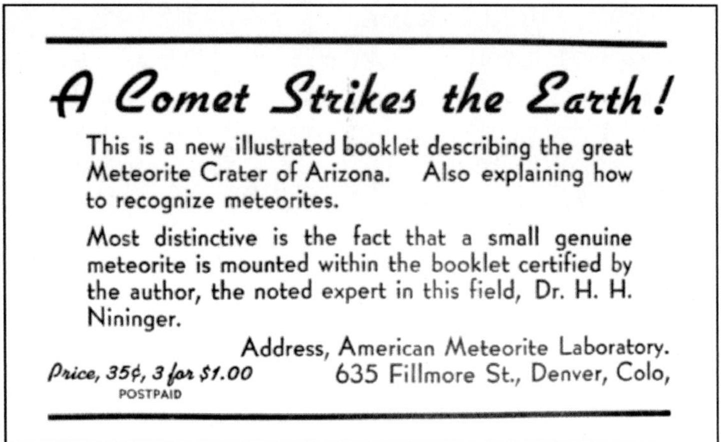
A Comet Strikes the Earth!

This is a new illustrated booklet describing the great Meteorite Crater of Arizona. Also explaining how to recognize meteorites.

Most distinctive is the fact that a small genuine meteorite is mounted within the booklet certified by the author, the noted expert in this field, Dr. H. H. Nininger.

Address, American Meteorite Laboratory.
635 Fillmore St., Denver, Colo,

Price, 35¢, 3 for $1.00
POSTPAID

Figure 3.21. *A Comet Strikes the Earth* promotional postcard that was addressed to Harry Locke of the Meteor Crater Observatory (discussed in Chapter 4). Courtesy Old Trails Museum/Winslow Historical Society.

The soon-to-be meteorite museum is not shown on the map. The most important change, however, was in the certification relating to the specimen of meteoritic material sold with the pamphlet. The original edition stated that the specimen was "a genuine meteorite." This edition changed the certification to read: "the attached specimen is an oxidized fragment of nickel-iron meteorite which was collected just outside the rim of Barringer Crater." The Society objected to the selling of specimens of meteoritic shale to the general public as genuine meteorites. In a review[40] of the revised edition, it was stated that "Surely no publication has profited more greatly in a scientific way by a single revisional change."

The pamphlet was revised again in 1951 to 69 pages and shows the American Meteorite Museum on the map page. The 1953 revision went to 76 pages and changed the publisher as the "American Meteorite Museum Sedona, Arizona." The 1953 version map moved the museum to Sedona. There was a substantial revision in 1969 after the museum closed. It was reduced to 66 pages and dropping the map. The publisher was the American Meteorite Laboratory, Denver, Colorado.

1947 *Chips from the Moon*

Chips From The Moon[41] (the title inside reads as *Chips from the Blasted Moon*) was published by the American Meteorite Museum, Winslow, in 1947. It is 40 pages, with black and white photographs of the Moon, varieties of tektites, and a map of worldwide tektite strewn fields. Tektites, are small black to green glassy objects, many of which are found over certain areas of the earth's surface, believed to have been formed as molten debris in meteorite impacts.

Figure 3.22. Cover of Nininger's "Chips from the Blasted Moon" pamphlet.

The booklet measures 6 x 4.25 in (about 11.0 x 15.0 cm). It deals with the mystery of tektites and their origins, specifically that they might have come from the Moon, blasted out by asteroids. His chief argument was that impacts must launch portions of the lunar surface at escape velocity; some of the debris must reach the earth. He argued that tektites are not found outside an equatorial belt some 90° in width. He considered that this would result from the position of the Moon's orbit. While this last suggestion has subsequently been shown to be inaccurate, Nininger's hypothesis was one of the important ideas of the present-day study of tektites.

1947-1948 Surveys[42]

Shortly after moving into the "Observatory" along Route 66 (discussed in Chapter 4), Nininger received a grant of $500 from the American Philosophical Society, plus funding from "two private donors," to continue the study of the distribution of meteoritic materials that began with the 1939 survey. Nininger obtained authorization from the Standard Iron Company, and from the Tremaine Cattle Company, who owned much of the land surrounding the Barringer property, to search for specimens. The October 6, 1947 letter from Brandon Barringer gave the authorization for six months and stated that "I see no reason to doubt that we will be able to arrange mutually satisfactory renewals from time to time."[43] Nininger was to supply a "report of the survey showing methods, area covered in detail, and finds by weight." The agreement also called for splitting any finds. Six months later, a second authorization, dated April 29, 1948, signed by D. Moreau Barringer, President, extended the permission through October 20, 1948.[44] The letter included the following tribute:

"I want to congratulate you on the most excellent article in the Saturday Evening Post. I think it pays only part of the tribute that is due your contributions to science, but it is a useful piece of publicity for your work as well as a very well deserved encomium." [The *Post* article as discussed in Chapter 1.]

Figure 3.23. Harvey Nininger with surplus metal detector searching for meteorites, from *The Nininger Collection,* Plate XXXVIII.

Nininger said that he had time in 1947 "to pursue the work at the crater under my permit." Even though the past four or five decades had supposedly stripped the area around the crater of meteorites, he was still determined to map the location of finds to assist in determining the direction of the meteorite at impact. Nininger had purchased a surplus metal detector and began the search. Figure 3.23 is an image of Nininger with the surplus Army metal detector on the plains around Meteor Crater.

In addition to the metal detectors, a survey using magnetometers began in the summer of 1948 to determine the distribution of large masses of at least 40 kg (88 lbs) which might be buried on the plains around the crater. He set up 1,032 stations, covering 23 acres along the base of the crater rim on the north and south, spaced 30 feet apart.

A handwritten note[45] of Nininger's, with the heading "Summary of Collecting on Barringer permits," is shown in Figure 3.24. He records collecting a total of 1,675 lbs (760 kg) of specimens, with 429.5 lbs (195 kg) delivered to Barringer in royalties as part of the permits.

Later, Nininger produced an undated map[46] (Figure 3.25) of the location of meteoritic materials. It appears this was created as part of his

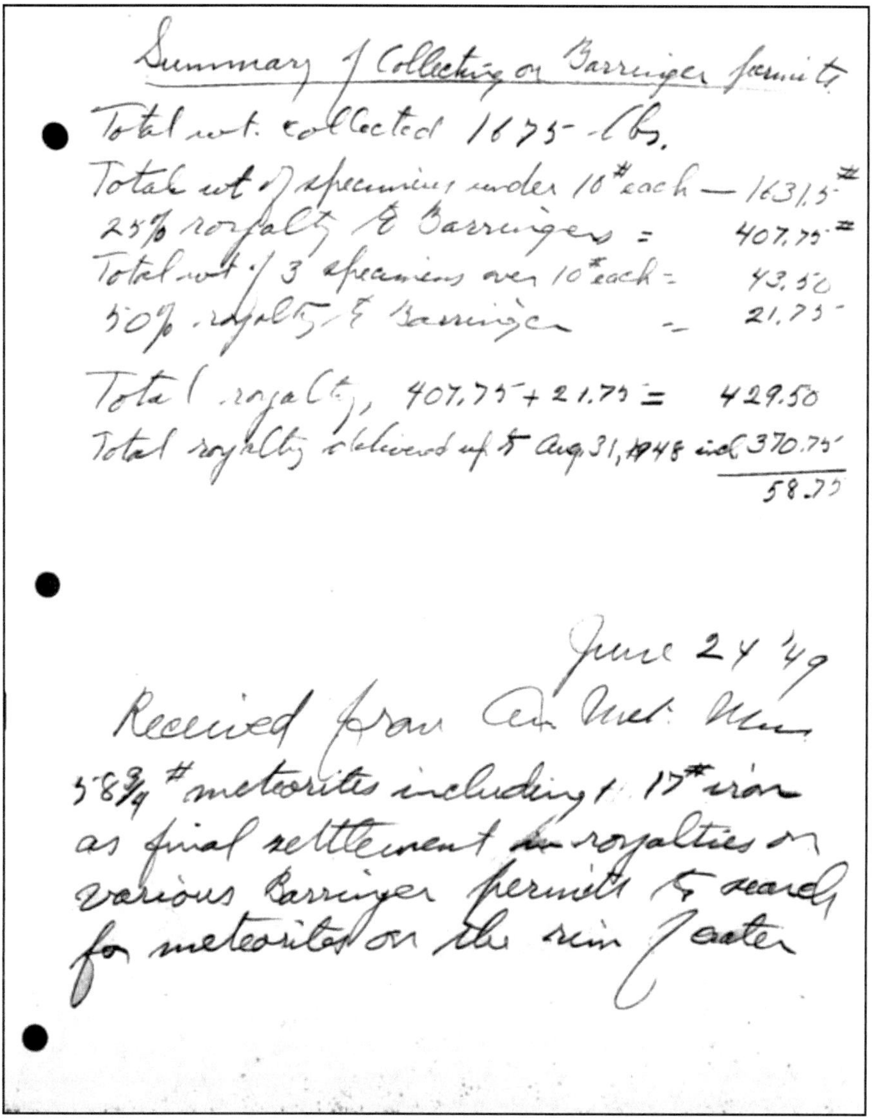

Summary of Collecting on Barringer permits

Total wt. collected 1675 lbs.

Total wt of specimens under 10# each — 1631.5#

25% royalty to Barringers = 407.75#

Total wt of 3 specimens over 10# each = 43.50

50% royalty to Barringer = 21.75

Total royalty, 407.75 + 21.75 = 429.50

Total royalty delivered up to Aug. 31, 1948 incl. 370.75

58.75

June 24 '49

Received from Am. Met. Mus. 58¾ # meteorites including 1 17# iron as final settlement on royalties on various Barringer permits to search for meteorites on the rim of Crater

Figure 3.24. Handwritten note from Nininger summarizing the collecting during the Barringer crater permits. Arizona State University, Hayden Library. H. H. Nininger Archives.

quantitative study to record the pattern of distribution and approximate weight of meteoritic materials. This data was used to develop his theory on the size and constitution (a large mass or a swarm) of the meteor, as well as the direction of its approach. He concluded that the meteor approached from the north-northwest as a large mass, accompanied by "a small number of lesser bodies." He suggested that as the mass approached the surface, the resulting shock shattered the mass, sending

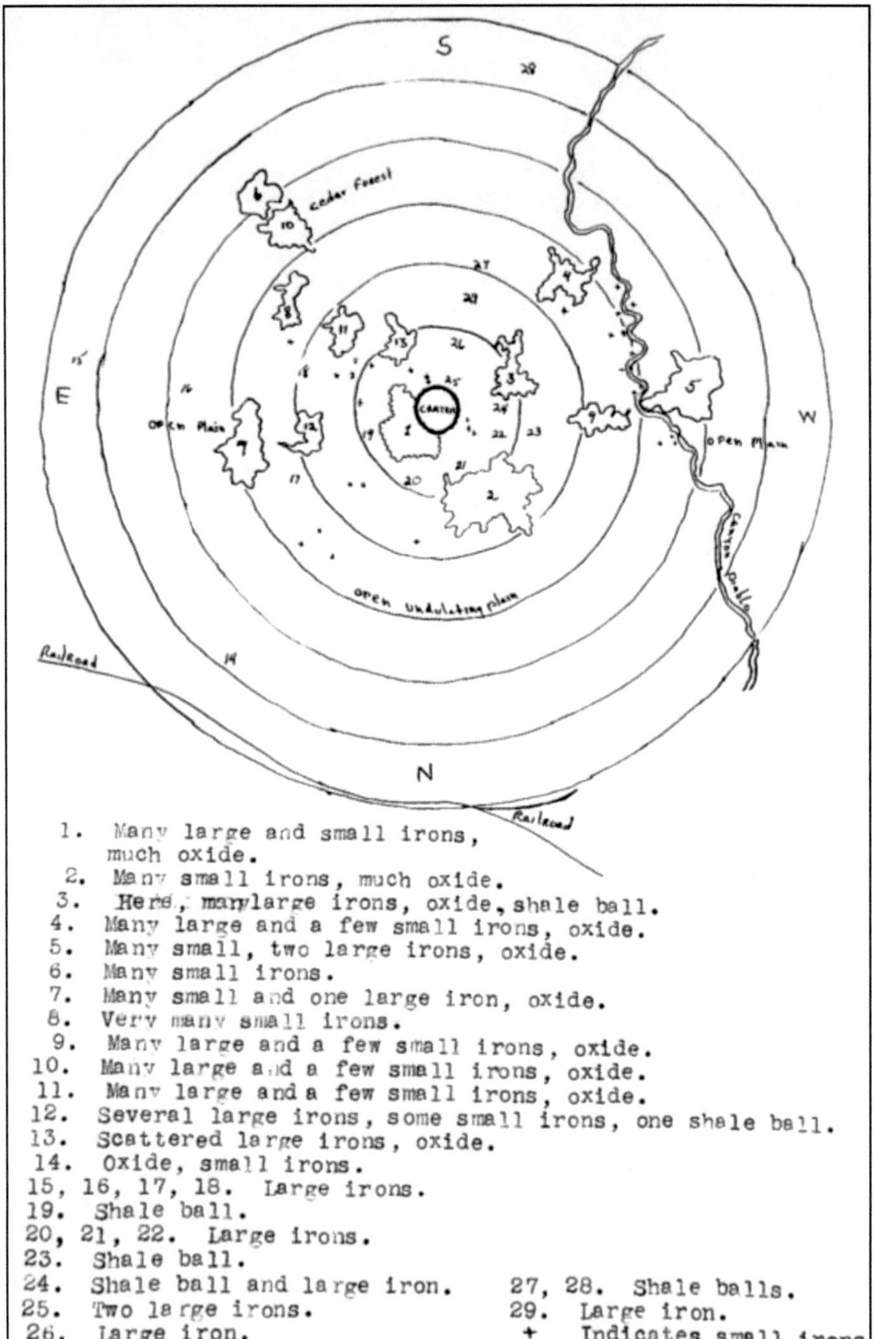

Figure 3.25. Nininger's map of the crater and location of meteoritic materials. Courtesy Arizona State University, Hayden Library, H. H. Nininger Archives.

fragments to spread "out over the surrounding plains." While some of these conclusions have subsequently been shown to be inaccurate, Nininger's contributions to study of the crater cannot be denied.

For reasons that are debated, during a 1948 meeting of the American Astronomical Society, Nininger presented a resolution calling for Meteor Crater to be put under state or federal control as a national park. The final portion of the resolution read:

> WE THEREFORE, The American Astronomical Society, in regular annual meeting assembled, do most respectfully suggest and request that proper steps be taken to induce a state or federal agency to acquire and set apart the crater, together with surrounding lands constituting an area of 100 square miles, more or less, with the crater as a public park to be preserved and operated in accordance with established laws and regulations governing such public parks.[46]

The resolution was passed. Nininger had not discussed this idea with the Barringers in advance. When they heard of it, the Barringers were already predisposed to not renew Nininger's permit because he was inviting anyone to walk the area to help collect material. They thought this violated the terms of the permit.

In addition, the Standard Iron Company had been approached by the Tremaine Cattle Company with the idea of constructing a visitor center on the rim of the crater. Architectural plans were submitted by Frank Lloyd Wright and Philip Johnson. Johnson's plans were accepted. Nininger's museum near the crater was seen as potential competition for this new endeavor.

Probably as a cumulative result of these factors, Nininger's permit to study the crater was not renewed after October 1948.

Spheroids — However, this did not stop Nininger from pursuing research with the huge amount of soil samples that he had accumulated at the crater going as far back as 1939. For example, Nininger began examining soil collected near the Meteor Crater rim during his 1946–1948

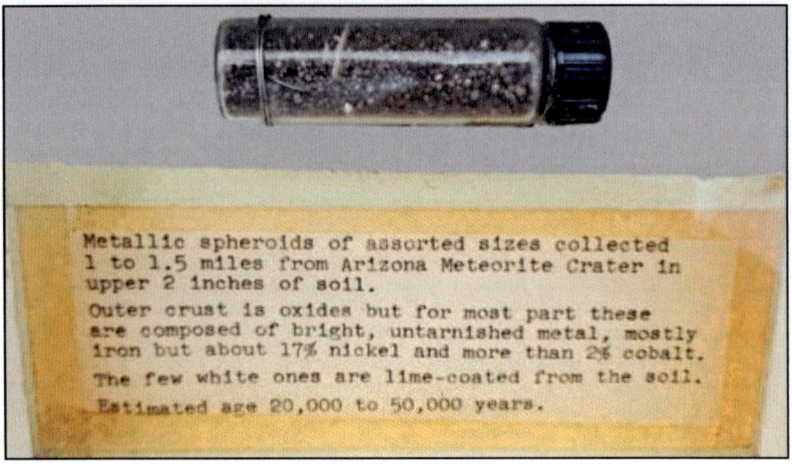

Figure 3.26. American Meteorite Museum exhibit of "metallic spheroids." Courtesy ASU Buseck Center for Meteorite Studies.

field investigation. The collected soil contained tiny spheres of nickel-iron that were the size of poppy seeds. In one of Nininger's notebooks, entitled "Notes on spheroid collecting," with a section heading of "Quantitative Studies of Droplets." He starts with "No. 101, coffee can full of magnetics from ant hills N.E. Crater rim." This continues through "No. 148." Figure 3.27 is one of the pages from the notebook.

When the large body that created Meteor Crater impacted the earth, it was theorized that the resulting explosion resulted in the majority of the body being vaporized. This would have created a huge cloud of hot iron-nickel vapor. As the cloud cooled, they would condense and fall in the form of iron-nickel rain.

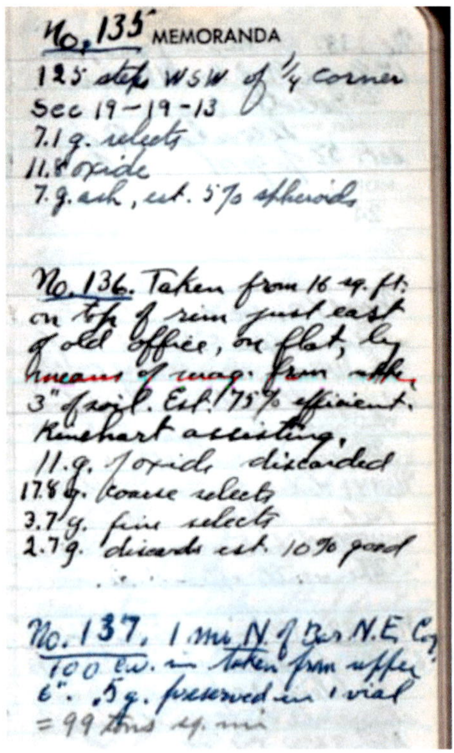

Figure 3.27. Page from Nininger's "Notes on spheroid collecting." Courtesy ASU Buseck Center for Meteorite Studies.

Nininger "worked a very long time at the problem of separating the little metallic spheroids" from volcanic ash, magnetic sand and other substances. He suggested that the nickel-rich metallic *spheroids* were condensation droplets from a cloud of vaporized metal created by the impacting meteorite. He believed that this constituted tangible proof of the explosion theory of how the crater was formed. In addition, he suggested that these tiny spheroids accounted for much of the mass of the meteorite that was not accounted for in the larger specimens that were found.

Figure 3.28. Nininger examining soil collected near the Meteor Crater rim during his 1946–1948 field investigation. Courtesy Dr. Carlton, Moore, ASU Buseck Center for Meteorite Studies.

Bomblets — Another Nininger discovery came when he was examining a gravel pit of material along the northwest crater rim that had been created by the state highway department. He discovered very small, tear-shaped melted objects, that he named ***bomblets***. They had never been described by earlier investigators, probably because they looked very much like the surrounding soil in which they were found. However, when he wet the pieces with water from his canteen, they revealed a dark greenish-gray color. When he ground some and examined it more closely, they revealed small imbedded metallic particles as "bright as chrome steel."

Nininger remembered seeing similar pieces on all of the crater slopes. He recalled that you could "kneel in a single spot and pick up 200 pieces without even moving." He then walked away from the crater and the number of bomblets thinned out rapidly beyond 1,500 feet from the crater, and almost none were found after one mile.

Other meteoritic craters that had been found around the world all had "impactites, " or tiny nickel-iron spheres formed by the melting of rock by the meteorite explosion. One of the "mysteries" of Meteor Crater had been why there were no such impactites there. Nininger believed the bomblets had landed while the debris of the lower Coconino rock layer was still falling from the meteor's impact. The bomblets were then covered by the falling rocks and missed being discovered earlier. He found the bomblets in large numbers, which solved this "mystery."

1952 *Out of the Sky*[47]

Out of the Sky: Introduction to Meteoritics was published in 1952 with a softcover version following in 1959. It was Nininger's hope that this book, on the subject of meteoritics, might be introduced in secondary schools as part of general science studies. He believed that it could also be used in college courses of geology and astronomy. As discussed in the first chapter, the subject of meteorites was not covered even in university courses as he was seeking this knowledge for himself. His hundreds of lectures and classes impressed upon him that people, with few resources, were interested in the subject.

Figure 3.29. 1952 *Out of the Sky* book dust jacket.

In the preface he thanks many institutions that provided him with access to their collections as he developed his own knowledge. He specifically thanked the Barringer family for granting him access to Meteor Crater over the years.

As the title suggests, the 336 pages, organized into 30 chapters, covers the types and composition of meteors, with discussions of several specific meteorite discoveries. It included 175 photographs on 52 full-page plates, plus 22 figures and illustrations. The book ends with his long-held proposal for a national institute of meteoritics, followed by a bibliography of his 124 articles and books from 1928 through 1951, many of which were used in this book.

1956 *Arizona's Meteorite Crater*[48]

Nininger's seminal work on his studies of Meteor Crater is presented in *Arizona's Meteorite Crater: Past, Present and Future.* Nininger, now 66 years old and in his Sedona museum, had completed his studies of Meteor Crater. In this book, he provided a thorough summary of the information that was available at that time on the crater. He describes the early history of the crater and outlines his theory on its formation. He covers the research and surveys that had been carried out, and critically reviews all the findings. He speculated on the nature of the event that took place when the meteorite encountered the earth. He ends with a list of 28 specific research projects that should be undertaken at the site. It is interesting that he chose to use the naming preferred by the Meteoritical Society, of "Meteorite Crater," rather than the popular "Meteor Crater."

Figure 3.30. 1956 *Arizona's Meteorite Crater* book dust jacket.

One reviewer found that "the book is well and interestingly written. It has both popular appeal and considerable scientific merit."[49]

Arizona Meteorite Crater and Me[50]

Within the volumes of material given to Arizona State University by Nininger, is a 20-page, undated, handwritten recollection of his work at Meteor Crater. The note includes background on establishing the museums near the Crater and in Sedona. The note includes a list of twelve discoveries that he made at the Crater. Yet, despite his years of work at Meteor Crater, Nininger recorded a situation in this note that disturbed him:

> "All of these discoveries and others have been described in my 1956 book *Arizona's Meteorite Crater*. Then imagine my shock that I experienced when several years later, on a visit to the crater, I picked up a beautiful little pamphlet beside the cash register, authored by Mr. George Foster, titled *The Meteor Crater Story*. I read it. In it were mentioned 36 scientists who were listed as men who contributed to man's knowledge of the crater. Some of these men I had accompanied on their first visit to the crater. For some unaccountable reason my name did not appear in the book. Some of my work is attributed to others and several other incorrect statements appear in the little volume."

The author of this 32-page booklet, published by Meteor Crater Enterprises in 1964, was George E. Foster. Foster was the Curator of the Museum of the Barringer Meteorite Crater that was established after Nininger moved his museum to Sedona. One can only speculate about this oversight. One possibility is that Foster had published a book in 1953 with the title "Arizona's Meteorite Crater," similar to the title of Nininger's book. The Foreword for Foster's book was written by Dr. Lincoln La Paz, of the Institute of Meteoritics at the University of New Mexico. An entire chapter could be written about the heated disagreements between La Paz and Nininger, so it is possible that La Paz influenced Foster to exclude Nininger from the list. Whatever the reason, this oversight was corrected in the 1996 edition of *The Meteor Crater Story*.

Figure 3.31. Harvey Nininger looking over Meteor Crater circa 1940. Courtesy of the American Meteorite Laboratory Photo Archives, Collections Research for Museums, Denver, Co.

CHAPTER 4

Showcasing the Collection

Harry Locke's Meteor Crater Observatory

Nininger's oil exploration job ended as World War II came to an end. After Nininger's friend and supporter J. D. Figgins retired as director of the Denver museum, the new director wanted the museum to go in a new direction with exhibits on birds and mammals. Although Nininger was invited to remain, the museum was not interested in further participation in his meteorite field work. After a four year hiatus of his meteorite investigations during the war years, Harvey and Addie decided that they needed to return to their independent ways.

They knew that they owned one of the world's largest private meteorite collection. Their goal had always been to increase the knowledge and research of meteorites. Since they were unable to obtain institutional support, they decided to open their own museum. After relocating to Flagstaff, Arizona, they began to look for a suitable location for their museum.

Nininger knew Harry Locke and his ill-fated Meteor Crater Observatory from his visit to the crater. He purchased some Canyon Diablo specimens from him that turned out to be his Canyon Diablo 2 discovery. He also mailed Locke a promotional postcard when *A Comet Strikes the Earth* was published.

Harry William Locke[1] was born in 1888 in Huntington, Oregon. After his service in World War I, he and his wife, Hope, traveled around the country and eventually found themselves at Meteor Crater in the 1920s. Locke was fascinated by the crater and soon became a fan and promoter of the crater.

They homesteaded 640 acres north of the crater at the intersection of Route 66 and the road to the crater. They called the intersection "Meteor Junction." Locke was a mechanic during World War I so he built a service station on a portion of the property. The Lockes opened "Meteor Station," a cafe and gas station, in the late 1920s, hoping the business would bring in enough money to fund their life's dream - the building of a meteor museum.

Figure 4.1. Harry Locke. Photo courtesy of the Old Trails Museum/Winslow Historical Society.

Locke sold gas at the station and would tell visitors about the crater. At the time, many visitors did not even know what a meteorite was, as Nininger had discovered. To help explain the crater and meteorites, he built a display in the station.

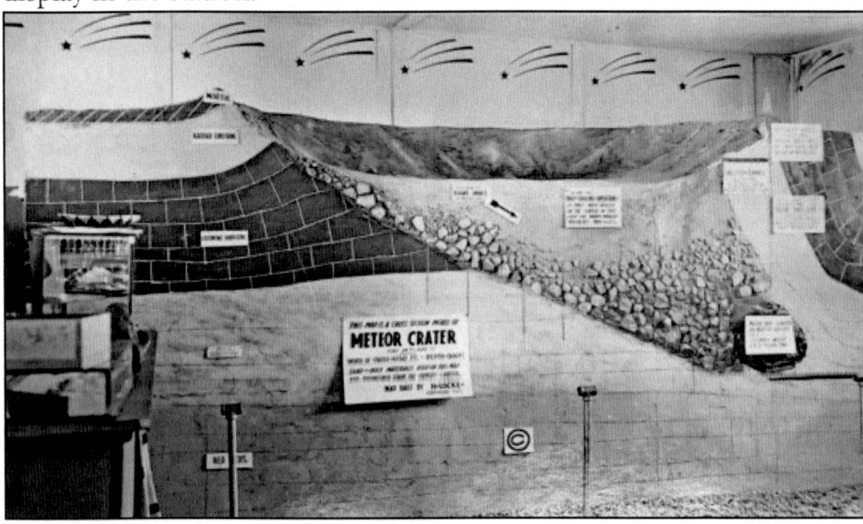

Figure 4.2. Meteor Crater display built by Harry Locke that was initially in Meteor Station and then moved to the Observatory. Photo courtesy of the Old Trails Museum/Winslow Historical Society.

Daniel Barringer died in 1929 but Locke was determined to keep his crater story alive. Needing funds to build a museum, he leased the Meteor Station to a colorful character by the name of "Rimmy Jim" Giddings for $30 a month plus $1,000 up front to cover rent for the last two years.[2] Rimmy Jim's became a landmark for travelers as shown in the area drawing in Figure 4.3.

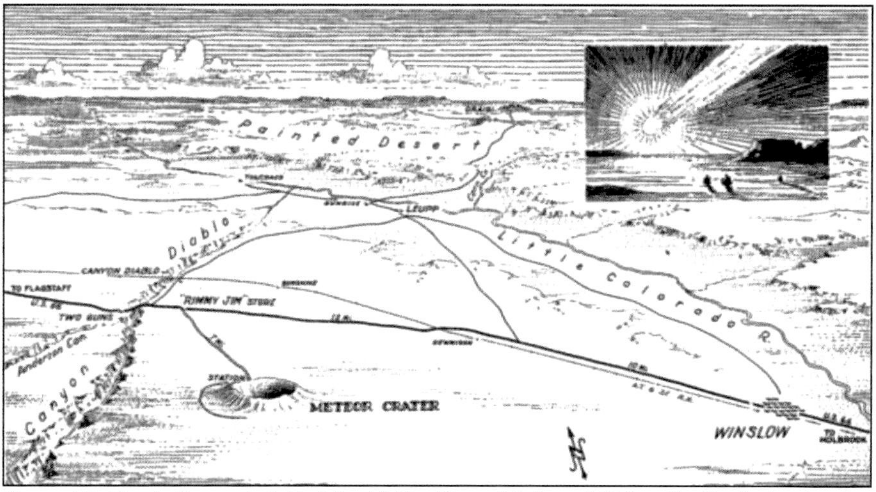

Figure 4.3. Map in the centerfold of the 1942 edition of *A Comet Strikes the Earth* showing the "Rimmy Jim" Store at the intersection of U.S. 66 and the road to Meteor Crater.

Using the lease advance, Locke started to build his "dream museum," with a viewing tower, south of the gas station. At that time, the crater was still privately owned by Barringer's Standard Iron Company, and tourists were not welcome on the active mining site. Locke went into the countryside and carved out the flagstone used in the building. He packed bricks together with a mortar mixture that he developed. He hoisted himself up in the air to complete the observatory tower. The floor was laid with flagstone and although it was a little rough, it was beautiful and practical.

Figure 4.4. Meteor Crater Observatory under construction circa 1936. Photo courtesy of the Old Trails Museum/Winslow Historical Society.

Figure 4.5. Meteor Crater Observatory opened in July 1936. Note the outhouse in the back. Photo courtesy of the Old Trails Museum/Winslow Historical Society.

The Meteor Crater Observatory opened in July of 1936 as a museum and rock shop with a tower where visitors could use a telescope to view the Barringer Meteor Crater. The Meteor Crater Observatory offered visitors narratives and models that described and illustrated how the meteor impacted the earth. As tourists would enter the building, they would find themselves in a large open space. On the wall near the door that lead to the observatory tower, was the relief display of the crater that Locke moved from Meteor Station. The museum had displays of Canyon Diablo meteorite fragments, and for a fee of $0.25, the visitor could go up the tower and view the crater through a telescope. A 1937 newspaper ad proclaimed ... "Continuous Lectures Pertaining to the Worlds Famous Meteor Crater of Arizona. Scientific, Educational, H. Locke, Lecturer." It was the culmination of the hard work of Harry Locke.

Through large picture windows, visitors could look west forty miles to see the San Francisco Peaks and a host of smaller purple hills and mountains. To the east from the tower, they could gaze toward the

Figure 4.6. Harry Locke business cards for the Observatory and for his cartoon business. Photos courtesy of the Old Trails Museum/Winslow Historical Society.

volcanic plugs and colorful mesas of the painted desert, thirty miles away. Locke would offer formal lectures to visitors.

To supplement his income he sold meteorite specimens and even made jewelry out of the meteorites. He developed a relationship with the Hayden Planetarium of New York, selling them Canyon Diablo meteorites. An order[3] from the Hayden Planetarium Sales Booth, dated March 22, 1937, acknowledged the receipt of twelve meteorite fragments and placing an order for "Twelve meteorite key rings." Business must have been good because in a letter[3] dated September 25, 1937, stated that "Our meteorite supply is getting low again, and I should like to have a shipment of small pieces whenever you can get them ready." Another letter, dated April 13, 1938, from the Hayden Planetarium reads: "The package of meteorites arrived today and we are very glad to have them.

Figure 4.7. Cover of Meteor Crater brochure produced by the Meteor Crater Observatory. Photo courtesy of the Old Trails Museum/Winslow Historical Society.

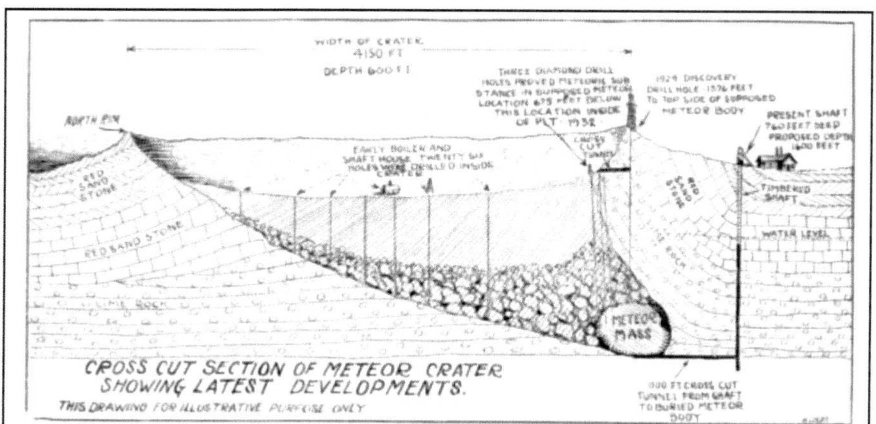

Figure 4.8. Page 5 of the Meteor Crater brochure with illustration by Harvey Locke. Photo courtesy of the Old Trails Museum/Winslow Historical Society.

Figure 4.9. Page 11 of the Meteor Crater brochure with cartoon by Harvey Locke. Photo courtesy of the Old Trails Museum/Winslow Historical Society.

I am sending under separate cover an order for the key rings . . . To date all the specimens have been sold."[5]

Figure 4.10. 1929 "Visitor Card" with meteorite specimen copyrighted by Harvey Locke. Photo courtesy of the Old Trails Museum/Winslow Historical Society.

As a cartoonist and illustrator, Locke created illustrations of the crater for displays and for the visitor brochure. This 12-page brochure, published in the early 1930s by the Meteor Crater Observatory, measured 6 in x 12.5 in (15.25 cm x 31.75 cm).

As noted in the correspondence with the Hayden Planetarium, Locke produced several items for sale to the tourists. One very interesting item was the sale of "Part of Heaven's Ancient Visitor" that included a meteorite fragment. Locke filed for copyright of this visitor card on January 15, 1929. He was selling this item in the Meteor Station before the Observatory was built. This may have been Nininger's inspiration to include a meteorite fragment in his publication of *A Comet Strikes the Earth*.

Unfortunately, the opening of the Observatory was not timed well as this was still during the Great Depression. Paying visitors were too few to

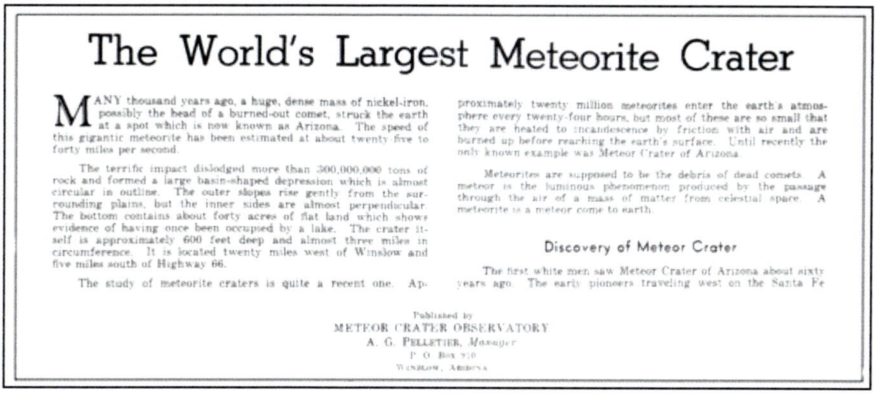

The World's Largest Meteorite Crater

MANY thousand years ago, a huge, dense mass of nickel-iron, possibly the head of a burned-out comet, struck the earth at a spot which is now known as Arizona. The speed of this gigantic meteorite has been estimated at about twenty-five to forty miles per second.

The terrific impact dislodged more than 300,000,000 tons of rock and formed a large basin-shaped depression which is almost circular in outline. The outer slopes rise gently from the surrounding plains, but the inner sides are almost perpendicular. The bottom contains about forty acres of flat land which shows evidence of having once been occupied by a lake. The crater itself is approximately 600 feet deep and almost three miles in circumference. It is located twenty miles west of Winslow and five miles south of Highway 66.

The study of meteorite craters is quite a recent one. Approximately twenty million meteorites enter the earth's atmosphere every twenty-four hours, but most of these are so small that they are heated to incandescence by friction with air and are burned up before reaching the earth's surface. Until recently the only known example was Meteor Crater of Arizona.

Meteorites are supposed to be the debris of dead comets. A meteor is the luminous phenomenon produced by the passage through the air of a mass of matter from celestial space. A meteorite is a meteor come to earth.

Discovery of Meteor Crater

The first white men saw Meteor Crater of Arizona about sixty years ago. The early pioneers traveling west on the Santa Fe

Published by
METEOR CRATER OBSERVATORY
A. G. PELLETIER, *Manager*
P. O. Box 710
WINSLOW, ARIZONA

Figure 4.11. Page 3 of the Meteor Crater brochure showing A.G. Pelletier as the manager of the Observatory. Photo courtesy of the Old Trails Museum/Winslow Historical Society.

support Locke's Observatory. The enterprise floundered and eventually closed with the building going into foreclosure. By 1938 it was listed for sale in the local newspaper: "10 acres on U. S. Highway 66 overlooking the famous Meteor Crater, San Francisco Peaks, and Indian country. Stone building suitable for cafe, club house, curio store. Grand location for cottage court."

The manager of the Observatory, A. G. Pelletier, purchased the property and changed it to an Indian Curio shop. They ran it for several years until they closed the shop in the winter of 1945-46.

American Meteorite Museum

Nininger had tried unsuccessfully to receive support from an institution that would support their collection. He decided that they would have to provide their own institution. They had longed to live near Meteor Crater for more than a decade and wanted to create "the first and only museum of its kind, and it would be run on a high, educational plane." In the summer of 1946, they moved their trailer to the base of Mt. Elden , east of Flagstaff, and began to look for a location for their museum.

Figure 4.12. American Meteorite Museum letterhead. Courtesy Arizona State University Hayden Library Archives.

Nininger[6] said that there were only two advantages of the Observatory building: it was situated on a small bluff just off the highway, looking across to the crater from a distance; and, it constituted a sort of landmark in itself. "It never was beautiful, at least not to my thinking, but it is an interesting structure, faced with natural flagstone, with a heavy square tower jutting against the sky." The fact that just a few miles to the south "rose the rugged rim of the most unique land feature known on this planet -- the great mile-wide meteorite crater . . . was our reason for choosing the location."

The remote location of the site was of some concern but they felt that they could do without police protection and "we could do without neighbors. We could get along without society and shows for entertainment. We could manage without church. Long since I had learned to derive more inspiration from Nature than the churches were giving me."

By the end of summer of 1946, they signed the lease and parked their trailer in the backyard. They arranged for expansion of the plumbing facilities in former storage space to accommodate visitors. The landlord dug a well and installed a water tank. The Niningers cleared everything out of the building, cleaned the picture windows and hung new venetian blinds across them. The walls were washed and painted. The living quarters consisted of one room about sixteen by eighteen feet, with a tiny adjacent bathroom. The floors were covered with sheet linoleum, "which rocked bumpily over the flagstone beneath." Book shelves separated the sleeping quarters from the kitchen and dining space. The cooking range was fed by butane gas from an outside tank. Since they did not have electricity, evening "found us in our single backroom, cooking and eating and reading by

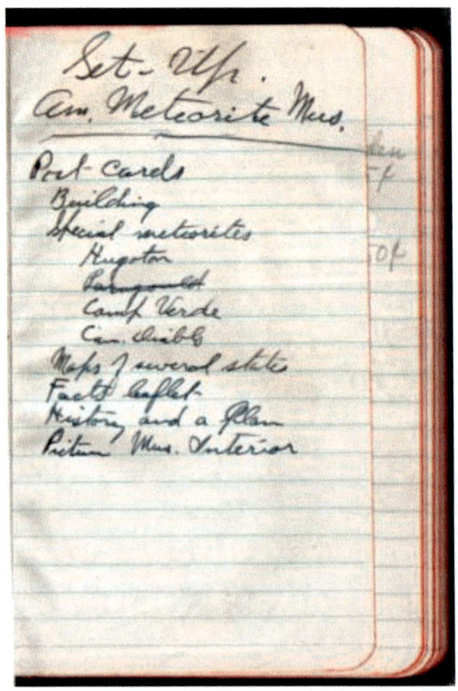
Set-Up.
Am. Meteorite Mus.
Post Cards
Building
Special meteorites
Hugoton
Camp Verde
Maps of several states
Facts leaflet
History and a plan
Picture Mus. Interior

Figure 4.13. Page found among Nininger's notebooks of to-do list for museum set-up. Courtesy ASU Buseck Center for Meteorite Studies.

kerosene or gas lantern . . . The only radio news we could get was received on the car radio."

With the building set up, they climbed into their 1942 Chevrolet and headed for Denver to bring back their collection of meteorites. The moving of a 16,000-pound collection of meteorites a distance of 750 miles was a large challenge. The packing required weeks of heavy labor. Eventually, however, all was accomplished, and the collection was loaded into the moving van. There were about a hundred boxes and crates, including two very heavy ones containing the two large stony meteorites, both of which were somewhat fragile. Hugoton weighed nearly 800 pounds before crating; Morland only a hundred pounds less. Both of these ranked among the half dozen largest stony meteorites in the world. There were eighteen heavy iron specimens, each weighing from 150 to more than 1,400 pounds. These were not crated and were in no danger of being damaged, but they constituted a real danger to everything else in the van in case of an accident, since any one of them, careening loose from a heavy jolt, could smash through any box or crate in the van.

Most important were the thousands of small specimens, some of which were worth several times their weight in gold, many of which were so fragile that they had been wrapped and labeled individually, in several thicknesses of tissue or toweling, before stowing them carefully in wooden boxes. The Denver museum was installing a series of new cases in the mineral department as the Niningers were leaving. The museum offered the old cases to Nininger, which he gratefully accepted.

Figure 4.14. American Meteorite Museum, circa 1949, with advertising billboards. Courtesy of the Sedona Historical Society.

The transfer agent contract contained a clause requiring the cargo to be locked into the van in Denver, and not to be opened until it reached the front door of the destination building. The company's estimate was that the truck would reach its destination twenty-four hours after departing, traveling day and night with two drivers. It did not arrive as expected at the end of the second or even the third day. The Niningers were starting to panic, but finally, the van arrived mid-afternoon of the fourth day. The

Figure 4.15. Museum interior photo of Harvey Nininger (right) with visitors. Courtesy of the American Meteorite Laboratory Photo Archives, Collections Research for Museums, Denver, CO.

Figure 4.16. Museum interior displays. Courtesy of the American Meteorite Laboratory Photo Archives, Collections Research for Museums, Denver, CO.

Figure 4.17. 1948 *Saturday Evening Post* photo caption: "Will I be hit? is what most visitors ask, eyeing the rocks that have fallen from the sky. The chances are slight - but there have been some near misses." Photos © SEPS licensed by Curtis Licensing Indianapolis, IN. All rights reserved.

Figure 4.18. 1948 *Saturday Evening Post* photo caption: "Look out below! This substantial monster is Doctor Nininger's prize. It flashed out of space, sometime, to bury itself near Hugoton, Kansas. Photos © SEPS licensed by Curtis Licensing Indianapolis, IN. All rights reserved.

Niningers, van crew and volunteers, worked until they had unloaded, and carried inside, about five tons of meteorites in 189 crates, bags and boxes along with the large display cases that the Denver museum had arranged for them to have for their specimens.

Most of their first winter was taken up with arranging and rearranging displays. All the big irons from Canyon Diablo had been placed on a central platform as the centerpiece for the room. As soon as a considerable part of the collection was housed under glass, they decided to open their doors. On October 19, 1946, the museum was opened. They gave three to five minute presentations daily, but spent more time with those showing more interest.

"Our hours were governed by the presence or absence of daylight, for as yet we had no electricity. As soon as light was good we were ready to open; as soon as good light was gone we closed, although on a few occasions we had such eager and persistent guests that we guided them about the cases by lantern light." The large display room of twenty by forty feet stretched across the front of the building. The tower, with some storage space to the side, opened at one end of the room.

Figure 4.19. Museum interior with Harvey Nininger (right) and school children. Courtesy of the American Meteorite Laboratory Photo Archives, Collections Research for Museums, Denver, CO.

As shown in Figure 4.20, they had developed ways to mount small meteorites on costume jewelry, which was added to their gift shop with the books and other items.

Nininger said that he was sure that during each year of the museum's operation, more hours of instructions were given to more people on the subject of meteorites than were given on this subject in any university. Nininger reported that the visitor count for November 1947 was over 1,000 and that during their first twelve months they had 33,000 paying visitors. Average daily visitors was between sixty to ninety.

During 1948, as compared to 1947, business all along Route 66 seemed to be generally down, but for the museum, on the whole, it was better, perhaps because they had established themselves as a going concern, or perhaps because 1948 brought unusually good press, with the museum featured in several magazines and newspapers, as discussed in Chapter 2.

Nininger lamented that along Route 66, with one "gyp joint" after another, with fake cave dwellings and Indian art, it was hard to convince visitors that they were not of the same ilk. The new Interstate highway had been finished by the end June 1949. Once it was opened to traffic, the museum noticed a drop in visitors and income by almost half. They were forced to lay off their two employees. From that time, the museum was largely run and managed by Harvey and Addie alone.

In 1950, Nininger recognized that their museum could not survive at this location. They attempted to sell their collection, but if they could not, "we must simply quit." Nininger sent a sixty-day notice to terminate the lease only to be told that they were bound by an automatic renewal provision of the lease, requiring them to pay the monthly rental, whether they occupied the building or not, until July 1, 1953.

They watched much of the traffic pass them by on the new Interstate highway. They talked of allowing the collection to go for a quarter of its value in return for a salaried post or an annuity on

AMERICAN METEORITE MUSEUM

PROUDLY PRESENTS

JEWELRY

FROM OUT OF THIS WORLD

☆

METEORITE JEWELRY

More rare than diamonds -- distinctively beautiful. No other precious or semi-precious gems can match the rugged gracefulness and individual beauty of these nickel-iron meteorite fragments polished along natural contours.

EARRINGS, button or pendant type, screw or earwire mounting of sterling (clip-ons, 50c extra)	$7.70
PENDANT, with 18" sterling chain	$11.00
pendant only	$9.35
BROOCHES, Bow Pin with pendant meteorite	$7.15
State of Arizona with "falling" meteorite	$11.00
BRACELETS, sterling, large set, Indian design	$33.00
sterling, 5 small sets, western design	$22.00
sterling, medium set, modernistic design	$14.30
CHARM BRACELETS, 5 meteorites, sterling chain	$14.30
1 meteorite, sterling chain	$4.13
charm only	$3.03
NECKLACE, 18", sterling, 6 meteorites	$27.50
CROSS, heavy sterling, 4 meteorites, 18" sterling chain	$16.50
RINGS, heavy sterling, large set	$14.30
small set	$12.10
CUFF LINKS, medium set, sterling stem	$8.80
small set, square rhodium plate back	$7.98
TIE CHAINS, adjustable	$7.70
TIE BARS, short, sterling, alligator type	$7.70
long, rhodium plate	$6.05
TIE TACS, small set on square of sterling	$5.50
large set	$7.15
KEY CHAINS	$5.50

ETCHED METEORITE JEWELRY

Polished and etched to reveal the distinctive Widmanstatten figure, these carefully shaped pieces of Xiquipilco, Mexico, meteorite are truly conversation pieces for the connoisseur.

EARRINGS, button or pendant type, screw or earwire mounting of sterling (clip-ons, 50c extra)	$19.80
CUFF LINKS, about 3-4" square, sterling stem	$35.75
TIE BARS, short, sterling, alligator type	$12.65
CUFF LINK-TIE BAR SETS	$46.00

LIST QUOTES MINIMUM PRICES INCLUDING 10 PER CENT FEDERAL TAX--QUOTATIONS ON LARGER ITEMS AVAILABLE UPON REQUEST.

Figure 4.21. American Meteorite Museum jewelry catalog. Note that items were made with pieces of the Xiquipilco, Mexico meteorite. Courtesy Arizona State University Hayden Library Archives.

Figure 4.20. Photo of jewelry from *Desert Magazine*, July 1942, with caption: "The seven meteorites in the necklace and the one on the watch charm are natural specimens. The others are polished specimens."

which they could live. Nininger made a rough estimate of the worth of the collection as $250,000, and on that basis made an offer to the Smithsonian Institution to sell the collection for $60,000. That offer stood until they decided in mid-1950 to build a new museum.

With the conditions of the lease requiring them to remain until July 1953, they had no other option, but to "tightened our belts." They cashed in some insurance policies and were able to at least break even until they could move to a new location.

Figure 4.22. Architectural modifications were made to the building, compared to Figure 4.4, such as adding windows to the observatory tower and front porch. A glass block entrance was added. The date of the modifications is unknown. Used with permission of the Newberry Library Chicago.

Sedona Meteorite Museum

They had been considering several western locations to purchase a homesite, and build a residence. In their search, they "had become entranced by the little town of Sedona," about 70 miles to the southwest

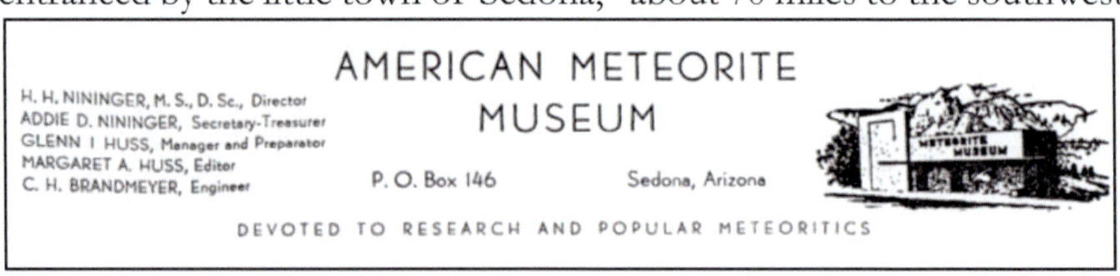

AMERICAN METEORITE MUSEUM

H. H. NININGER, M. S., D. Sc., Director
ADDIE D. NININGER, Secretary-Treasurer
GLENN I HUSS, Manager and Preparator
MARGARET A. HUSS, Editor
C. H. BRANDMEYER, Engineer

P. O. Box 146 Sedona, Arizona

DEVOTED TO RESEARCH AND POPULAR METEORITICS

Figure 4.23. American Meteorite Museum in Sedona letterhead. Courtesy Arizona State University Hayden Library Archives.

of the "observatory" location. They decided to build a combination museum and living quarters.

In *Find a Falling Star*, Nininger explains that the decision to move and build a new museum in Sedona was "on the assumption this beautiful little town would become a mecca for nature-minded tourists. The move was good; the town boomed; but the boom was in terms of retired couples and art-oriented people, mostly . . . We assumed that as the town grew, and awareness of our museum's location grew, business would

Figure 4.24. Construction of the Sedona Meteorite Museum begins in the summer of 1953. Courtesy Sedona Historical Society.

Figure 4.25. Construction of the Sedona Meteorite Museum continues in the summer of 1953. Courtesy Sedona Historical Society.

increase, but it stayed about the same." There were few structures and paved roads, or electricity in parts of Sedona in the 1950s. An groundwater aquifer under West Sedona was not discovered until 1951, which only then opened the way for development in that area. The real boom in Sedona tourism did not come until the late 1960s. The city was not officially incorporated until 1988. However, the Niningers had some "new friends" who offered to loan them funds to build a new building to meet their needs in Sedona. Perhaps this funding source was an added incentive to move to Sedona.

Figure 4.26. Sedona Meteorite Museum front entrance. The building was completed in September 1953. The sign to the left of the door reads: "The New Home World Famous Nininger Meteorite Collection." Courtesy of the American Meteorite Laboratory Photo Archives, Collections Research for Museums, Denver, CO.

The first few months were successful which made them feel that the decision to move was correct. In 1955, daughter Margaret and her husband Glenn Huss, came to Arizona. Glenn was offered the job of Manager and Margaret provided helped with the visitors. They both assisted Nininger as he finished his book, *Arizona's Meteorite Crater*, and with several revisions to *A Comet Strikes the Earth*.

However, as the years went by it became more and more difficult. Not yet a "tourist town" by the fall of 1957, it became obvious that the winter months were never going to be good, and the summer months were not successful enough to carry them through the following winter. They expected the town to "boom" but the boom did not bring the type of people interested in their museum. Once again they had to face the question of whether to keep the museum open or sell the collection.

Figure 4.27. Sedona museum interior. Courtesy of the American Meteorite Laboratory Photo Archives, Collections Research for Museums, Denver, CO.

Figure 4.28. Sedona museum interior with Nininger (right) talking to a customer. Courtesy of the American Meteorite Laboratory Photo Archives, Collections Research for Museums, Denver, CO.

Figure 4.29. Sedona museum interior gift shop with Addie Nininger behind the counter. Courtesy of the American Meteorite Laboratory Photo Archives, Collections Research for Museums, Denver, CO.

Figure 4.30. Sedona museum exterior window display (Figure 4.26) with telescope. The telescope was donated to McPherson College by granddaughter Peggy Schaller. Courtesy of the American Meteorite Laboratory Photo Archives, Collections Research for Museums, Denver, CO.

Figure 4.31 The Sedona Meteorite Museum included living quarters in the back, lower level of the building. Courtesy Sedona Historical Society.

Figure 4.32. The Sedona Meteorite Museum poster. Courtesy Sedona Historical Society.

The Buildings Today

The American Meteorite Museum along Route 66 was abandoned and eventually sold to the Bar T Bar Ranch Company. All that remains are some remnants of the building and tower.

The Sedona Meteorite Museum was eventually sold and is now part of the Sedona Arroyo Roble Hotel & Creekside Villas.

Figure 4.33. The American Meteorite Museum along Route 66 as of 2010. © Greg Disch Photography with permission.

Figure 4.34. Sedona Meteorite Museum on State Route 89A in Uptown Sedona, 2020.

CHAPTER 5

Sedona to Retirement

He has been, and is, an independent scholar resisting the trend of conformity leading to mediocrity.

McPherson College, 1976 Ceremony

1958 British Museum of Natural History

As time in Sedona went by, the boom in tourists that they were expecting never materialized. Most new residents were either retiring or were in to the art scene. Visitation was steady but never increased to make it profitable.

The curator of meteorites for the British Museum of Natural History had indicated that they were interested in purchasing a portion of the collection for display and research. In a letter dated October 11, 1957, Nininger offered the British Museum 276 specimens from the 680 falls in the collection. The specimens were selected to ensure that the value of the remaining portion of the collection was not impacted. The British Museum found the offer acceptable but needed time to raise funds.

In the meantime, as word of the potential sale spread, interest in the collection was shown by the Smithsonian Institution and by Arizona State University. However, neither institution had the funds at the time to make a firm offer. In a letter, dated June 13, 1958, the British Museum made a firm offer for 1,200 specimens, or about 21 percent of the Nininger Collection. The offer was accepted and packing and shipping of the specimens began.

The contract provided for reimbursement for the "loss" of weight during the cutting, grinding and polishing of some of the specimens. Since

Nininger had left the large saw in Denver, he used a laboratory in California for some of the work. In a "Balance Sheet"[1] note, Nininger recorded four shipments to the British Museum beginning on July 15, 1958. The bulk of the collection shipped on October 10, 1958. The total value was shown as $158,766 (equivalent of $1,625,905 in 2023 dollars[2]).

1958 - 1959 Orient Expeditions

The sale allowed the Niningers to pay off their debts and leave a sufficient amount to continue operation of the museum. Daughter Margaret and son-in-law Glenn Huss, moved into the lower living quarters of the museum. In addition, the sale provided funds to allow the Niningers to take a 72-day expedition to Australia, New Zealand, Fiji and Hawaii. With the bulk of the meteorites shipped to the British Museum, they sailed from San Francisco on October 26, 1958, leaving Margaret and Glenn to run the museum.

One of the purposes of the trip was to study craters and search for meteorites and tektites. They were accompanied by friend, Allan Kelly. Kelly was born on December 5, 1900. Kelly was a rancher and an avocational geologist who spent decades conducting research into impact geology. He probably came to Nininger's attention with the 1944 publication of an essay, "A New Theory of the Flood," in which Kelly suggested that geologic change could be explained by the impact of meteorites onto the surface of the earth. Over the next 50 years, he refined his hypothesis that cosmic collision was the determining factor in the physical and biological development of the earth. The first and most comprehensive statement of his ideas came in *Target: Earth*, a book that he published in 1953 with chemical engineer Frank Dachille.[3] At the time of its publication, Nininger noted that it was largely ignored by the geological community, yet he considered it "a great book that will mark the beginning of a new epoch in the study of geology."[4]

Keeping abreast of the meteoritic literature, Nininger was aware of the Dalgaranga crater in Western Australia. In December 1938, a brief account of a meteorite crater located on the Dalgaranga Sheep Station, was published,[4] together with the description of a 1.5 oz (42 g) nickel-iron meteorite fragment which reportedly came from its vicinity. The author of the paper had not visited the site, but rather relied on reports of the former manager of the Dalgaranga Station. The dimensions were given as 230 ft (70 m) in diameter and 15 ft (4.6 m) in depth.

Nininger was interested in the explosive formation of meteorite craters, such as Meteor Crater and the Odessa Crater, both discussed in Chapter 2. Nininger believed that a relatively small crater, such as Dalgaranga, could exhibit evidence of its formation by the explosive vaporization of the meteor, as had occurred at the Meteor and Odessa craters. One diagnostic piece of evidence that could be used to test whether this occurred was the presence of metallic tektites (Chapter 3), which formed when the meteorite melted on impact and solidified as it cooled. Nininger believed[6] a crater the size of Dalgaranga could shed light on the issue.

He would need to visit Australia to prove his theory. Such a trip, however, would have been prohibitively expensive until 1958, when he sold part of his meteorite collection to the British Museum.

They flew into Perth, Australia, on January 31, 1959.[7] On February 1, with directions from some locals, they drove to the Dalgaranga Crater but needed the help of the "Station manager, and an aborigine employee" to guide them to the site. They reached the crater on February 9 and set up camp near the crater rim.

They soon realized that the crater measurements in the 1938 study were in error. In his notebook[8] on this trip, he wrote: "Crater is nearly round and gives one the impression of a miniature C.D. [Canyon Diablo] as to relation of width to depth." Kelly and Nininger produced drawings of the crater (Figure 5.1).[9]

Unfortunately, the metal detector they brought to search for meteorites would not work properly so they had to look for meteorites visually. They also searched for tektites using magnets. The notebook records that:

> "Allan picked up 10 and I 16 small irons around this crater. Mostly they were found on the side where the rim seemed to have blown away. Allan turned 9 of his over to me, keeping the largest for a charm to be made. "

This is slightly different than what was presented in the published report[10] in which he noted: "There was a notable scarcity of meteorite specimens, and in two days the three of us recovered only 23 small irons with an aggregate weight of 149 g." Nininger was also frustrated that they, nor any of the locals, had found any tektites. His notebook includes an interesting observation:

> "We have noted a very good reason why tektites could very well be present here and still not be collected. The Kangaroo droppings when fresh are so like the lens type tektites in size, shape and color that one is continually stepping on them to test their true nature when looking for the tektites."

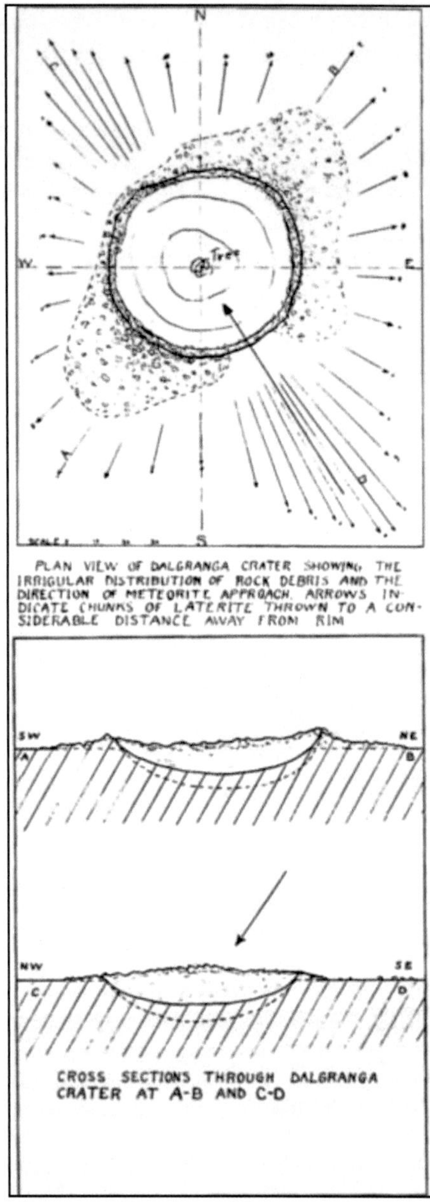

Figure 5.1. Drawings of Dalgranga Crater made by Kelly.

This presents an amusing picture of Nininger going around, stepping on Kangaroo droppings. He does not carry this observation into the published paper, but there is a similar passage in the paper: "Laterite pebbles constituted a considerable nuisance throughout our investigation. They not only had almost the exact colour of meteorites. . . . "[11] He may have realized later that Kangaroo droppings and laterite pebbles look very similar.

Nininger thought that the crater appeared to be an explosion impact crater caused by the violent fragmentation of the meteorite and not by vaporization. However, the low number of fragments collected seemed to contradict the fragmentation hypothesis. He reasoned that perhaps one reason was that the meteorite was not a nickle-iron meteorite but that it may have been a stony-iron. Due to the lack of time, they concentrated on collecting the material for future analysis. They mapped the crater and the distribution of meteorites before leaving early the next afternoon to visit the other objectives of the trip. The answer to "the Dalgaranga puzzle" would need to wait until he processed the materials collected back home.

Nininger, Addie and Kelly left Dalgaranga Crater after two days to visit the other known Australian meteorite craters (fifteen in three locations) that were on their itinerary. They visited Henbury craters where there are 13 to 14 craters ranging from 23 to 591 ft (7 to 180 m) in diameter and up to 49 ft (15 m) in depth that were formed when the meteor broke up before impact. Several tons of iron-nickel fragments have been recovered from the site. Next was the Boxhole Crater, the site of nickel-bearing metallic fragments and iron shale-balls similar to those found at Henbury. Lastly, they visited the Wolfe Creek Crater that averages 2,871 ft (875 m) in diameter, 200 ft (60 m) from rim to present crater floor.

Sedona Historical Landmark Home

Before they left on their trip, the Niningers arranged to purchase a hilltop lot in Sedona. They hired Howard Madole (1923-2015) to design the house. They received and approved the plans while on the trip. Madole had arrived in Arizona in 1948 when his parents purchased a 10-acre lot in West Sedona. He had no formal architectural training. His degree from the University of Iowa was in science and commerce. However, he designed and built his parent's home on their lot. Madole continued to learn the construction business and worked on several projects. By 1952, he had the largest construction company in Northern Arizona.[12]

Madole was involved with Frank Lloyd Wright at Taliesin West where he helped to build the Usonian design structure there. He was a frequent dinner guest of Wright. Madole went on to develop his own style with signature elements, many of which were included in the Nininger home.

While the house was under construction, they lived in their travel trailer. The home was eventually completed so that they could move in by Thanksgiving Day of 1959. It was one of the first homes built in the subdivision, so the road was named Meteor Drive.

Figure 5.2. City of Sedona Landmark plaque on the Nininger house on Meteor Drive.

Figure 5.3. Nininger house on Meteor Drive showing the Madole's signature cantilevered roof extensions.

Figure 5.4. Nininger house showing the Madole's signature windows that "die" into the fireplace.

The 1,250 sq.ft. home has many signature Madole features, such as of a 2 x 4 wood ceiling/roof, laminated wood beams that support the roof with large cantilevered roof extensions, local stone fireplace with picture windows that "die" into the fireplace and planter penetrated walls, visible both in and out. The house was added to the list of Sedona Historic Landmarks in 2009.[13]

1959 Dalgaranga Expedition

Upon examining the Dalgaranga material back in Sedona, more questions were raised then answered. Nininger decided that he needed to return to Australia for a more complete investigation. He suspected that the lack of meteorite fragments may have been due to larger masses having been buried that avoided surface collecting. He planned to extend his surface search and to excavate within the crater. The Museum of Western Australia endorsed his plan.[14]

Nininger and Glenn Huss arrived at Dalgaranga on October 18, 1959. They set up camp on the southwest rim of the crater. On their first visit, only a rough crest-to-crest measurement was taken, but on the second visit a complete measurement was taken. The next 12 days involved almost continuous searching with at least one instrument, as well as visual searching. During that time, only 207 specimens, with a combined weight of 2.4 lbs (1.1kg), were recovered. Excavations within the crater produced more specimens, but no large mass.

Figure 5.5. Figure 5 from the 1960 Nininger-Huss paper showing Dalgaranga crater looking northeast across the crater, showing the greatest aggregation of rubble, mainly laterite.

Their report stated that: "The crater, measured from crest to crest of the crater rim, was found to have a diameter of about 70 ft. Its depth we found to be only about 10.5 ft."[15] Nininger concluded that Dalgaranga was created by a large mass, composed mostly of stone but containing some stony-iron materials. It struck the earth at sufficient speed to disintegrate on impact. The stony constituents were reduced to dust and have been scattered and carried away by the forces of erosion.

It was recognized, however, that after the crater's discovery, 40 years earlier, it was finally surveyed, and the true nature of the Dalgaranga impactor was reported by Nininger in the scientific literature for the first time.[16]

1960 - Arizona State University Sale

Upon his return, Nininger came to the realization that the Sedona Museum could not maintain itself. In the fall of 1959, Arizona State University (ASU) again indicated that they were interested in the remaining items in the collection. Several other institutions expressed interest, but no firm offers were received. An inventory was taken to determine its value, to be ready for any sale.

The convergence of two events helped lead Arizona State University toward meteorite research. First, Sputnik launched in October 1957, and put space and anything related to it at the forefront of the American conscience. Second, as described above, Nininger had sold a portion of his collection to the British Natural History Museum in 1958.[17]

In 1958, Arizona State College became Arizona State University (ASU). In an effort to bolster the research program of the new university, the University Coordinator of Research, George A. Boyd, sought promising research avenues. Boyd was familiar with Nininger's collection and recognized its importance to Arizona and to ASU's pursuit of research in an up-and-coming discipline. The British Museum sale marked a loss for the State of Arizona but added an urgency on the part of ASU to acquire the balance of the Nininger Collection.

In a letter[18] to George Boyd, dated December 9, 1959, Nininger made the following offer to ASU:

> "Pursuant to our telephone conversation this morning, I am enclosing a nearly completed catalog of our collection which we are offering to Arizona at a price of $275,000 [$2,816,245 in 2023 dollars[19]]. The material included on the enclosed list invoices in excess of $500,000.00. We are offering all of the specimens that are now exhibited in our museum and quite a number that are not on exhibition. The sale price also covers 26 falls which have not yet been catalogued or invoiced and one fall the purchase of which is still being negotiated. When this cataloging is completed the collection will number 684 falls."

The letter concludes with the following:

> "Besides the 684 falls represented, the offering includes generous quantities of the little metallic spheroids and impactite discovered by Nininger at the Arizona crater, and other materials of a strictly scientific meteoritical nature which offer fertile ground for research.
>
> Two cutting machines, one for stony and one for metallic meteorites will go with the collection, together with a complete catalog of all items at no extra cost."

The Niningers wanted to attend a conference on meteoritics in Copenhagen in August 1960 so it was decided to close the museum in May 1960 and begin packing the collection if a sale came through. They left for Copenhagen in July 1960 and toured Europe for four months, visiting universities and meteorite collections. At about the same time, Margaret and Glenn moved back to Denver to carry on the work of the American Meteorite Laboratory.

To acquire the collection, ASU applied for a grant from the National Science Foundation (NSF), with supporting funds from both the ASU Foundation and from Mr. Herbert G. Fales, the vice president of International Nickel Company, who was familiar with Nininger through his own interest in meteorites (Chapter 3). The NSF also recognized the importance of keeping the remainder of Nininger's collection in the United States and accepted the ASU proposal on June 8, 1960. Upon the Nininger's return from Europe, shipping the collection to ASU began.

With the Nininger collection en route to ASU, a search was begun for a director of the newly established Center for Meteorite Studies. They consulted with Harrison S. Brown, a geochemistry professor at the California Institute of Technology, who was one of the few scientists in the nation actively studying meteorites, to find a worthy candidate. Brown was also familiar with Nininger and his collection; he had obtained samples for study from Nininger and had also visited Nininger's museums with his students. Brown recommended one of those students, Carleton B. Moore, for the directorship. Moore's PhD thesis sought chemical groupings of chondrite meteorites and ultimately helped improve the understanding of chondrite chemical abundances. Acting on behalf of ASU, Herbert Fales flew to Wesleyan University where Moore was teaching at the time, to recruit him. Moore agreed to take the position.

Figure 5.6. Carleton Moore studying meteorite samples. Photo courtesy of Carleton Moore.

The initial activities of the Center for Meteorite Studies (CMS), began in the spring of 1961. Moore's first task as Director was to organize a symposium on meteorite research, which brought together scientists from a wide variety of fields. The proceedings of the conference were published as "Researches on Meteorites," which provided a valuable snapshot of the state-of-the-art knowledge of meteorites at that time. Concurrently, the first Advisory Board met to help formulate the operational guidelines for the CMS. Brown again aided ASU by serving on the Advisory Board, alongside other scientific luminaries. It is a tribute to the first Advisory Board that much of their input has stood the test of time. For instance, the Center's current scientific loan policy is largely based on the original recommendations of the Advisory Board.

The Advisory Board also emphasized the directive from the NSF that the collection be "available primarily for the purposes of scientific research by qualified scientists throughout the History of the Center." In the formative years of the CMS, Moore acted on this direction in a number of ways. The Center began loaning collection specimens to qualified scientists for study and even secured a NASA grant to keep the loan service free of charge. Moore, himself, increased the visibility of the Center by serving as Editor of *Meteoritics*, the journal of The Meteoritical Society, and arranging for ASU to publish the journal.

In conjunction with its expanded scientific activities, the Center began its foray into educational and public outreach activities with the opening of the Center for Meteorite Studies Museum in 1967.

Figure 5.7. Arizona State University's Buseck Center for Meteorite Studies in 2022.

1960 American Meteorite Laboratory

The Sedona museum was closed and Margaret and Glenn were back in Denver to carry on Nininger's work by reestablishing the American

Meteorite Laboratory. The era of Addie and Harvey Nininger's American Meteorite Museum had ended.

AMERICAN METEORITE LABORATORY

P. O. BOX 2098 -:- DENVER 1, COLORADO

GLENN I. HUSS, Director
MARGARET A. HUSS, Secretary

H. H. NININGER, Consultant

Figure 5.8. Letterhead of the American Meteorite Laboratory circa 1960. Courtesy of the Sedona Historical Society.

The Laboratory was opened on July 1, 1960. In the announcement of the opening, Glenn Huss stated:

> The American Meteorite Laboratory will:
>
> 1. Carry on a continuous search for new meteorites through the free examination of suspected meteorite samples, conducting field investigations where indicated and purchasing authentic meteorites at fair prices.
> 2. Distribute to scientists, institutions, and collectors meteorite specimens, books on meteorites, and a variety of other materials covering all phases of the subject of meteoritics. Catalogs of these materials, as well as of our unique meteorite jewelry, are available upon request.
> 3. Promote our nation's defense and conquest of space by furthering in every way possible the recovery and study of meteorites and related phenomena.

The Laboratory operated until Glenn's death in 1991. Margaret kept the Laboratory minimally open, but it was formally closed in 1998 due to Margaret's declining health. Margaret passed away in 2007. Glenn Huss made a significant contribution in his own right by having collected an additional 239 meteorites that were previously unknown.

1961 - Ask a Question About Meteorites

As seen from his travels, retirement was not in Harvey Nininger's vocabulary. In this 1961 publication written by Nininger, and published by the American Meteorite Laboratory, he continues his lifelong quest to educate the public on the subject of meteorites. *Ask a Question About Meteorites*[20] is an 87-page booklet, of which 32 pages are photographs. It is in his familiar 4.5 x 6 in. format. In the six chapters, he posed questions with answers that relate to each subject. This represented the best knowledge of meteorites that was known at that time. Below are the chapters with a sample question and answer that is still applicable today.

Figure 5.9. Cover of *Ask A Question About Meteorites*.

Meteors and Meteorites - 35 questions

Why do meteorites fall?

The use of the term "fall" in relation to the arrival of a meteorite upon the earth is somewhat deceptive. In space there is no "up or "down." Each body in the solar system, whether large or small, travels its own particular orbit operating under the influence of the larger body of which it is a satellite.The earth travels in its orbit around the sun at a speed of about 18.5 miles per second. Meteorites, traveling around the sun in somewhat more elliptical orbits, move at speeds averaging about 26.2 miles per second in the neighborhood of the earth. If a meteorite's orbit brings it too close to the earth, the earth's gravitational pull diverts it from its course and it collides with the earth. However, the term "fall' is both descriptive and convenient. It is therefore common for both the scientist and the layman to say that a meteorite "falls" to the earth and to refer to the total amount of material which comes down from one fireball as a "meteorite fall."

The Nature of Meteorites - 36 questions

What is the best test that one who is not a scientist can make on a suspected stone to determine whether it is a meteorite?

Take the stone, or a chip from it, to an emery wheel and grind into it slightly. Then examine the ground surface for a content of bright, silver-looking metal. If it is a stony meteorite the metal will appear in the form of irregular grains, usually very small. If it is a metallic meteorite its interior will appear like a piece of steel. If either of these results is obtained then a nickel test is in order. This any good chemist can make.

Classification of Meteorites - 13 questions

What is meant by etching?

The etching of a meteorite is the application of a weak acid solution to the polished surface of its metallic portions in such manner as to cause the crystalline structure to become visible. Etching is possible because in the crystallization process the metal of a meteorite has separated into two principal alloys - one of low nickle content and one of high nickel content. When the acid is applied it dissolves the different alloys at different rates producing etching pits of various depths in the polished metal.

Meteoric Clouds and Dust - 20 questions

What are meteoritic clouds?

Many of the meteorites which have been seen to fall have left in their wakes clouds of dust or "smoke." These clouds, which are composed in part of dust from the meteorite, may appear roundish or as long ribbon-like clouds.

Meteor Craters - 23 questions

What is a meteorite crater?

A meteorite crater is a bowl-like depression in the earth which is formed by the impact of a large meteorite. It will tend to be circular in shape (although some are elliptical), the pit floor is lower than the surrounding terrain, the pit is surrounded by a prominent collar or uplifted rim, the strata in this rim dip sharply away from the vertical axis of the pit, and there is no evidence of extrusion of lava in any of the known craters.

Tektites - 6 questions

What are tektites?

Tektites are the small, dark glass objects which are found distributed over several large areas of the earth's surface and which seem to bear no relationship to the terrains on which they are found. The name tektite is derived from the Greek word *tektos*, meaning molten, and was chosen because of unmistakable evidence that these glassy particles were shaped while in plastic condition.

A 1962 review[21] noted that "The standard terminology approved by the International Astronomical Union is not followed in the text. However, these small points do not decrease the value of this pamphlet as a guide to meteorites."

1962 Nininger Meteorite Award

At a 1962 conference on meteorites, Harvey and Addie announced the establishment of a $25,000 [$246,697 in 2023 dollars[22]] endowment fund to promote interest in meteorite-related topics among young scientists. The Nininger Science of Meteoritics Fund at the Buseck Center for Meteorite Studies at Arizona State University, supports the Nininger Meteorite Award, which recognizes outstanding student achievement in the "Science of Meteoritics" as embodied by an original research paper. Applications are considered by an Independent review panel of field experts. The Nininger Award recipient receives $2,000, and an engraved plaque commemorating the honor. Nininger recounted that "Our greatest satisfaction came from seeing the awakening realization among scientists that the general field of meteoritics is basically important, that certain aspects of this young science have vital bearing on problems that have plagued geologists and astronomers."[23]

Figure 5.10. The Nininger Science of Meteoritics Fund is administered by the Buseck Center for Meteorite Studies.

1966 Niningerite

Recognition and honors were finally coming to Harvey Nininger. In 1966, he was honored to have a newly discovered mineral named after him. The new iron-magnesium-manganese sulfide mineral, was discovered by Dr. Klaus Keil, research scientist at the National Aeronautics and Space Administration's Ames Research Center. **Niningerite** is an opaque grayish mineral, having a metallic luster. It occurs as microscopic crystals, and it is generally found in a certain class of chondrite meteorite.

Figure 5.11. Example of niningerite. Courtesy www.assignmentpoint.com.

Dr. Keil wrote to Nininger that "We propose naming the new mineral *niningerite* in recognition of your outstanding contributions to the field of meteoritics." The name and definition was approved by the International Mineralogical Association's Committee on New Minerals and New Mineral Names.[24]

1969 Meteorite Crater Study Kit

Margaret and Glenn Huss continued to work with Nininger on various projects and publications through the American Meteorite Laboratory in Denver. They worked with Nininger to produce the 5th and last version of *A Comet Strikes the Earth*. This final version, copyright 1969, was 65 pages in length, with 20 images, and an oxidized fragment of a nickel-iron meteorite that was attached, similar to all the previous version. The centerfold map of meteor crater and the museums was not included.

Comparing the last two editions, the 4th edition from 1953 was 76 pages in length, with 23 images, an attached oxidized fragment, with the centerfold map of Meteor Crater. Most of the pages not included in the 1969 edition relate to Meteor Crater and to the "puzzling pits" from the TWA expedition (Chapter 2).

In addition, a *Meteorite Crater Study Kit* (Figure 5.12A)[25] was developed and released by the American Meteorite Laboratory in 1969. Contained within a 4.75 x 6.5 x .75 in. box, the kit included the following items:

- Leaflet entitled *Sparks from a Celestial Bomb* with several "nickel-steel particles" (Figure 5.12B). The interior of the cover showed a drawing of Meteor Crater. The two pages of text, entitled "A Rain of Nickel-Steel," is a brief history of the crater describing the contributions of Barringer, Moulton and Nininger.
- Meteorite explanatory card with an attached fragment (Figure 5.12C). The card explains that the sample is 90 percent iron and 7 percent nickel, with minor amounts of other metals.
- Impactite explanatory card with two attached specimens (Figure 5.12D). The card explains how impactites are formed and that thousands of tons of these "little bombs" escaped discovery until Nininger discovered them in 1953.
- The 1969 revised, *A Comet Strikes the Earth* (Figure 5.12E).

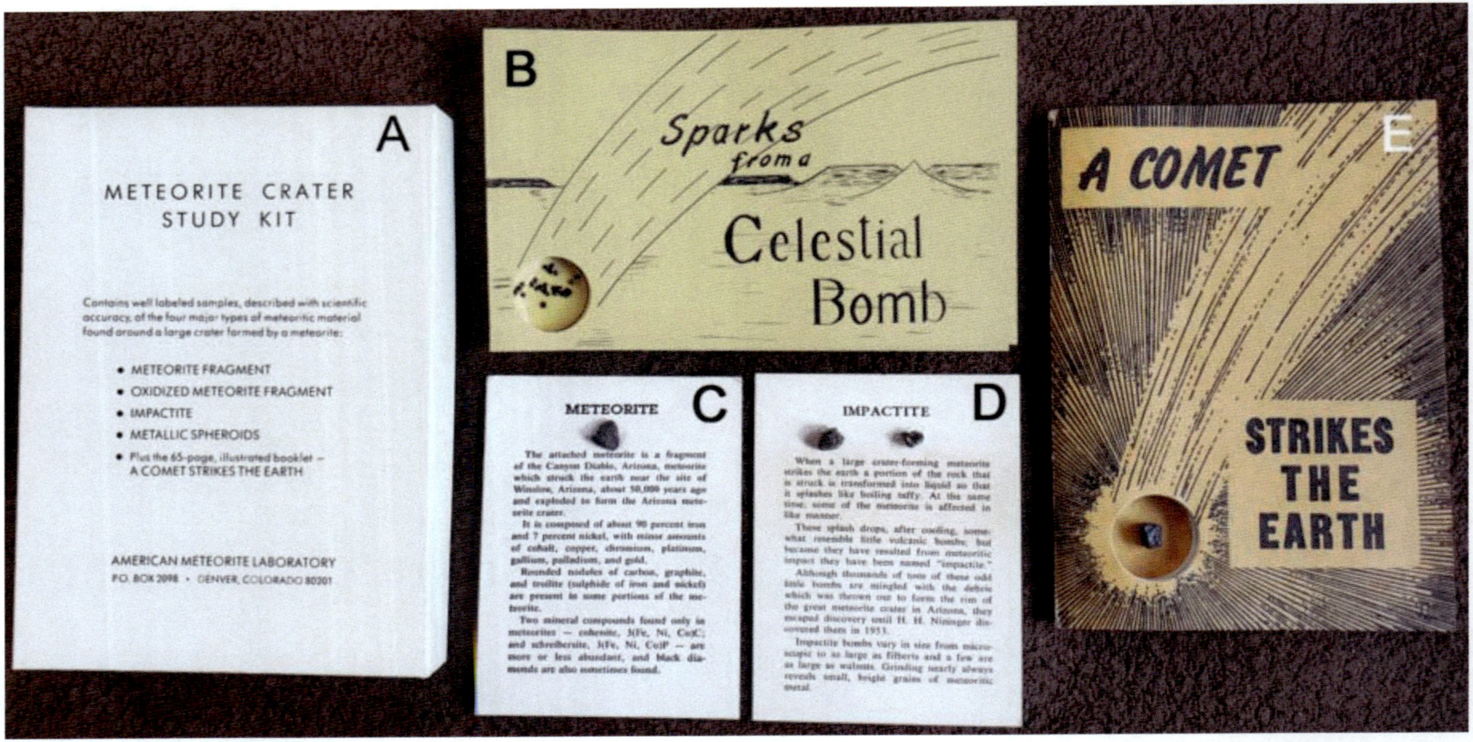

Figure 5.12. The Meteorite Crater Study Kit (A) contained a pamphlet entitled *Sparks from a Celestial Bomb* (B), a card with a *Meteorite* (C), a card with an *Impactite* (D), and a copy of *A Comet Strikes the Earth* (E).

It was stated that all of the specimens attached to the cards were collected from Meteor Crater.

1967 McPherson College Honor

Nininger received his Bachelor of Science degree from McPherson College in 1914. In 1967, the College presented Nininger with an honorary Sc.D. The presentation ceremony recognized his meteorite collection and the wealth of his books and publications. The most poignant portion of the presentation noted the following:

> "These scientific accomplishments mark him as a world renowned meteoriticist and qualify him for this honor tonight. He also is honored for a virtue of even greater value. He has been an individualist following his conviction of the necessity for collecting meteorites in anticipation of the Space Age. This vision often lacked recognition from his peers, and was carried on without institution embrace and without government support. He has been, and is, an independent scholar resisting the trend of conformity leading to mediocrity. His faith in the value of his search on behalf of mankind carried him through forty years."[26]

This author could not have summarized his amazing career and personality more eloquently.

1972 - Find a Falling Star

The starting point for many of the details in this biography of Harvey Nininger has been his 1972 autobiography entitled *Find a Falling Star*.[27] The published edition had 254 pages, with 24 black & white photos. The original manuscript was nearly twice as long as the published version. His publisher, however, suggested that he needed to shorten it. This must have been a difficult task, but it helps to explain why some important discoveries were not included in the final version, and why, in some cases, descriptions are cut short. However, accuracy and completeness of this biography was assured by access to the original manuscript at Arizona State University's archives of the Nininger papers.

The published version of Harvey H. Nininger's autobiography provides a fascinating read into his accomplishments, along with some of his frustrations, as well as their financial difficulties and struggles. Book reviews, at the time, must have been gratifying to Nininger after so many years of skepticism and disregard.

Figure 5.13. Cover jacket of *Find A Falling Star*.

> "Seldom in the history of science has one person had so large an influence in developing a field as H. H. Nininger. Now we have the opportunity to follow Nininger's experiences through his new book. The vivid descriptions and details of his past meteorite searches are brought forth with freshness and clarity to the reader of this fascinating autobiography. It is a tribute to Nininger that he has found time from his busy schedule to document and publish his innumerable adventures and bitter disappointments. The experiences described by Nininger during his 80+ years are considerably more than most people will ever have."[28]

Another review at the time found that "All in all this is a fascinating book describing a unique style of work in the spirit of the nineteenth-century entrepreneurs."[29]

Although out of print, some copies are available through antiquarian booksellers who also provide complimentary reviews.

> "A meteorite hunter's bible! Harvey's thrilling autobiography,

published in 1972, is beloved by meteorite fanatics the world over. Long out-of-print and difficult to find. A must-have for meteorite collectors, and meteorite hunters."[30]

"In this remarkable book, Dr. H.H. Nininger often called the world's foremost meteorite expert, recounts the excitement of his long love affair with 'falling stars'- those 'stones from the sky' that rain down upon our earth from outer space in great, but little known profusion."[31]

"This is the classic on meteorites by the famous Meteor Crater scientist who started the meteorite craze. A Cult classic and great reading."[32]

"Seminal meteorite hunter Harvey H. Nininger's extraordinary life was recounted in his thrilling autobiography, *Find a Falling Star* — a must-read for all meteorite enthusiasts."[33]

1976 Pomona College Honor

Nininger received his Masters degree in Biology at Pomona College in Claremont, California, in 1917. President David Alexander sent a letter dated April 5, 1976, to Nininger with the following honor:

"On the recommendation of the Faculty, and the Committee on Honorary Degrees, the Board of Trustees has elected you to the honorary degree of Doctor of Science in Pomona College. This honor, the highest in the gift of the College, is richly deserved by you for your contribution to the advancement of understanding about the solar system and your indefatigable study of meteorites which places you among the important scientists of the twentieth century."[34]

The honorary degree was conferred upon him at the Commencement Ceremonies of June 6, 1976.

1978 Addie Nininger

It has been clear throughout this book, that Nancy Adeline "Addie" Nininger was an integral part of Harvey's meteorite career. They were married for 64 years, and had three children, Robert D. (1919-2004),

Doris Elaine Nininger Banks (1921-2012), and Margaret Ann Nininger Huss (1925-2007). Addie ran the household and managed their finances, both at home and during their extended field trips, including the early "house car" trips. She acted as Harvey's field assistant, helped to establish and run both of the meteorite museums, and was the co-author of *The Nininger Collection of Meteorites*.

Figure 5.14. *Life* magazine, September 6, 1937, with the caption: "Addie Delp Nininger accompanies her husband on most meteorite-gathering field expeditions. She cooks, tends their trailer home, is also an expert meteoriticist in her own name." (c) Life Picture Collection/Shutterstock.

Addie died in 1978 at the age of 86. In Harvey's autobiography *Find A Falling Star*, he recognized the contributions of Addie by noting:

> "I doubt that the world holds another woman who would put up with such demands as my program placed upon Addie. When she accompanied me on field trips, it was not merely for the outing. She helped with the driving, with field notes; and, whenever we were using a trailer, kept house and meals in readiness just as at home, while I dashed about after all sources of information, scurried off on strenuous field work or delivered a lecture or several lectures a day. She expanded her activities as the children grew."

1982 Minor Planet 2421 Nininger

According to the International Astronomical Union (IAU), a minor planet is an astronomical object in direct orbit around the Sun that is neither a planet nor a comet. Minor planets include asteroids. Minor planets observed over at least two nights, and which cannot be identified with an existing celestial object, are initially assigned provisional designations. When enough observations of the same object are obtained to calculate a reliable orbit, a number is assigned to the minor-planet designation. After the designation is assigned, the discoverer is given an opportunity to propose a name, which, if accepted by the IAU, replaces the provisional designation.

A minor planet was discovered from the Lowell Observatory, Anderson Mesa Station, on October 17, 1979 by Edward L. G. Bowell.[35] He was principal investigator of the Lowell Observatory Near-Earth-Object Search. He has discovered a large number of asteroids. Bowell chose to name this object *Nininger* on the occasion of Harvey Nininger's 95th birthday, on January 17, 1982.

Return to Colorado

After Addie's death, Nininger continued to live a full life. He remarried Gladys L. (Griffin) Dawson (1903-2000) on March 28, 1979. In 1981 Nininger moved back to Colorado. The Nininger home in Sedona was passed on to their daughter Doris (1921-2012).

He attended the Meteoritical Society's 50th anniversary meeting in Mainz, Germany in 1983. Harvey died in Westmister, Colorado, after a short illness on March 1, 1986, at age 99. He passed away shortly after the return of Halley's Comet on February 9, 1986, which he had been looking forward to.

Figure 5.15. A happy Harvey Nininger, surrounded by his "friends." Courtesy Sedona Historical Society.

Epilogue

This biography provides a brief overview of Harvey H. Nininger's life. For those in the field of meteoritics, or who have read his autobiography, *Find a Falling Star*, many of the basic facts of his life, such as education, work, relationships, and accomplishments, may already be known. The use of images from a variety of sources, assembled in one place, hopefully adds to a better understanding of his fascinating life.

Some biographies include an analysis of the subject's personality. Editorial reviewers of this book used descriptions such as "obsessed" in describing Nininger. An obsession, by definition, is something that excessively preoccupies the mind. This is not necessarily a negative assessment, but an equally descriptive term for his chosen field could be "passion." While these terms seems to apply, perhaps a better insight can be obtained from the following undated paper, handwritten by Nininger, found within the *H. H. Nininger Papers* in the Archives at Arizona State University.

Reasons For Giving Meteorite Investigations Top Priority In My Life

When my interest in meteorites was aroused by the great fireball of November 9, 1923, I began to search for sources of additional information on the subject. I met with disappointment in every institution to which I turned except the Field Museum in Chicago, the U.S. National Museum and Harvard University.

Even in the 3 institutions they seemed to be content with the status quo, except for an eagerness to acquire a sample of each new meteorite. I found no one concerned about the matter of how much meteorite material arrived on the earth and no one seemed to doubt that the collections in museums of the world constituted an adequate qualitative sample of meteoritic increment. Papers on the overall composition of meteorites were common and the quantitative aspects of the subject were considered unimportant.

The general attitude in the scientific world appeared to be fairly represented by a member of my audience when I addressed the Cosmos Club in Washington, D.C., in the early '30s. At the close of my lecture, opportunity was given for remarks or questions and a middle-aged man of eminence spoke: "And now young man, since you have learned how to find meteorites and you have learned all these things about them, of what use is it?"

To my way of thinking he may as well have asked the same question of the paleontologists who were searching for and describing new species of fossils, or the archeologists who were tracing out the forbearers of mankind. Remains of ape-men were scarce indeed but their importance was not measured in terms of the number of recoveries. Geologists had said that meteorites were quantitatively of no importance; but I didn't believe it, neither did I believe that man yet knew all of the kinds of matter that reached the earth from space. The little more than a century since scientists had first admitted their existence was too brief a moment in the long geological development of our planet to decide that meteorites were not a factor in the planet's development, in my opinion.

The rarity of falls, the fact that almost no community could be found wherein a fall had occurred within memory of the living made very difficult an estimate of the frequency of falls notwithstanding there may well have been falls in each community since the last ice age. And then there was the matter of large falls, such as produced the Arizona crater and larger. Records were not wanting of mountain-sized asteroids straying into the earth's domain. Within 4 years after I had first suggested earth-asteroid collisions, the Hermes object, a mass of billions of tons, passed too near the earth for comfort, if only the people had been aware of it. Passage within the earth's orbital domain had been occurring about once in a decade and it seemed quite unreasonable to assume that during a half billion years all such intruders had missed our planet.

Geology texts all contained references to geological revolutions during which major crustal deformations had taken place over much of the earth's surface. To my way of thinking asteroidal collisions could best account for such changes, along with polar shifts, continental drift, etc.

To me it was also inconceivable that man's very brief glimpse at the earth's cosmic relations could possibly have given him a sufficient basis on which to catalog the different varieties of meteoritic increment.

I realized that many of these ideas were too far out for me to hope that an objective consideration would be given them at the time, if indeed, I should live to see them seriously considered, and I reasoned that pressing for their acceptance would only alienate me from the main-stream of geological thought. Consequently I pursued my search for meteorites knowing that by this means, I was gathering concrete, visible evidence that my subject was of some importance, that its importance had been sadly overlooked and that sooner or later those other aspects of the problem would come under critical scrutiny; then if I were right, a favorable verdict would be rendered by the geological and astronomical communities.

Unfortunately, as he laments above, he lived in an era when most scientists ridiculed those who recovered, studied, and researched meteorites. It was then considered of little or no scientific use. Harvey Nininger was the pioneer of modern meteorite research, and many consider him the father of contemporary meteoritics. He led the way when others thought the science was frivolous. His collecting methods, cataloging, and displaying of meteorites are in many ways still the standard today.

As discussed in the preceeding chapters, Nininger was an exceptionally successful meteorite hunter. He collected specimens of 226 meteorite falls not previously identified, of which only eight were actually seen to have fallen. He began collecting and studying meteorites during a period when the leading scientists of the day thought there was nothing left to learn. At the time, many thought that looking for meteorites was a waste of time because they believed that they were so uncommon. He recovered a substantial portion of the meteorites that are available to scientists today. By the time Nininger retired, the Space Age had dawned and meteorites had become a key component of research into the origin of the Earth and the Solar System.

Glossary

This book is not intended to be a primer on the study of meteorites, however, the use of terms used in meteoritics is unavoidable. The following list of terms may be useful. However, several texts (in no particular order) on the subject are recommended for those looking for more in depth information.

Field Guide to Meteors and Meteorites by O. Richard Norton and Lawrence Chitwood

On the Trail of Stardust: The Guide to Finding Micrometeorites: Tools, Techniques, and Identification by Jon Larsen

Meteorites: The Story of Our Solar System by Caroline Smith, Sara Russell and Gretchen Benedix

Meteorite: How Stones from Outer Space Made Our World by Tim Gregory

---

Aerolites - Meteorites are generally classified into two main groups, according to their chemical composition. The aerolites, or stones, contain some 20% of iron, but their principal constituent is silica, which forms about 40%. (See also Siderites.)

Asteroid - small planet-like bodies that orbit the sun lying mostly in the region between the orbits of Mars and Jupiter.

Chondrites - A large sub-class of stony meteorites are chondrites. They are so named because of the presence of numerous granular inclusions, or chondrules, which are commonly of a glassy nature. Chondrites are primitive aggregates of early Solar System materials.

Class / Classification - is a meteorite's scientific designation into a particular type of specimen. (e.g. L6, EH5, etc.)

Comet - a body which orbits the sun and is primarily composed of frozen water, ammonia, methane, carbon dioxide and many pieces of rock and dust. Their distinct tail occurs when they approach the sun from their suspected origin in the outer edges of our Solar System.

Crater - formed from the collision of a large natural body from space when it impacts a large body's (eg. planets, moons) surface at high velocity.

Fall - a meteorite which was witnessed to fall to Earth's surface.

Find - a meteorite which has been found and has no record of being witnessed to fall to Earth's surface.

Fireball - a very large and bright meteor which can also often leave a smoke trail.

Fusion Crust - in the last couple of seconds of luminous flight, the molten surface of a meteor solidifies to form a thin, usually black, fusion crust.

Impactite - If a meteorite strike is large enough to form a crater, it radically deforms the rocks, sand, or soil around the point of impact. The terrestrial materials, physically altered by the tremendous heat, pressure and shock waves associated with very large meteorite impacts, are known as impactites.

Iron (Siderite) meteorite - meteorites composed mainly of iron and nickel.

Meteor - is a meteoroid that has entered the Earth's atmosphere and burns brightly leaving a flaming tail and sometimes smoke. They are more commonly called 'Shooting Stars' or 'Falling Stars'.

Meteorite - is a meteor that has survived entry through the atmosphere and reached the Earth's surface.

Meteoritics - the study of meteorites and associated phenomena.

Meteoroid - Meteoroids are objects in space that range in size from dust grains to small asteroids. When meteoroids enter Earth's atmosphere at high speed and burn up, the fireballs or "shooting stars" are called meteors. The progression of terms starts with a *Meteoroid* in space that becomes a *Meteor* when it enters the Earth's atmosphere, and then becomes a *Meteorite* if it survives entry and lands on the ground.

Pallasites - Pallasites is a type of stony-iron meteorites. Only about 0.3% of known meteorites are pallasites. Pallasites come from the boundary between the metallic core and the rocky mantle of an asteroid.

Siderites - Meteorites are generally classified into two main groups, according to their chemical composition. The siderites, or irons, are composed predominantly of iron, with the addition of some 10% of nickel, and rather less than 1 % of cobalt. (See also Aerolites.)

Siderolites - Meteorites of intermediate composition are sometimes called siderolites.

Stony (Aerolite) meteorites - composed predominantly of silicate minerals.

Stony-Iron meteorites - composed of iron-nickel metal (metallic iron) and silicate minerals in roughly equal proportions. Stony-iron meteorites are typically about 50:50 stone:metal.

Strewn field - the usually elliptical area over which fragments from a shower of meteorites are distributed on the ground.

Tektite - Tektites are small, pebble-like glassy objects of Earth material that have been melted by meteorite impact, splashed up into our atmosphere, and fallen back to Earth. They often acquire aerodynamic shapes when they partially melt during their fall to Earth.

Widmanstätten pattern - Iron meteorites are from the metallic core of asteroids. The iron cools so slowly that large grains of interlocking metal form. If an iron meteorite is sawn, polished, and etched with an appropriate weak acid solution, the individual metal grains are revealed in a geometric pattern called a Widmanstätten Pattern. Widmanstätten patterns do not occur in stony meteorites. They are only seen in iron meteorites and pallasites.

NOTES

Chapter 1

1 Nininger, H.H. Interview by Susan L. Rogers, 28 January 1976. Tape recording, Flagstaff Public Library Oral History Project.
2 Miller, Arthur M. 1923. "Meteorites." *The Scientific Monthly* 17(5): 435.
3 Unless otherwise noted, quotes are from Nininger, H.H. *Find A Falling Star*. Paul S. Eriksson Inc. New York, 1972.
4 Nininger, H. H. "Visitor From a Distant Planet." *Desert Magazine*, July 1942.
5 Nininger had originally written his autobiography, *Find A Falling Star,* in a much longer version than what was published. The publisher believed that the original version, in excess of 500 pages, would be difficult to sell so requested Nininger to cut it back to what was eventually published. James Banks, grandson of Nininger, provided sections of the original version to help clarify some sections in the published version. Other original draft sections were obtained from the Arizona State University Archives.
6 Typed remarks by Addie Nininger for an address titled "Gypsying" to the student body at Chapel, McPherson College, 1925-1926. On file at the Sedona Historical Society in binders provided by Harvey Nininger.
7 Nininger, H.H. 1933. "Meteor Craters vs Steam Blowouts." *The Mines Magazine* 23(12).
8 Nininger, H.H. 1935. "How to Recognize a Meteorite." *The Mines Magazine* 25(1).
9 Clarke, R. S., H. Plotkin. 2002. "UCLA Astronomer Frederick Charles Leonard (1896-1960): From Childhood Prodigy to Mature Obsession." American Astronomical Society, 201st AAS Meeting, id.37.05; *Bulletin of the American Astronomical Society*, 34: 1159.
10 Nininger, H.H. "When the Sky Rains Stone and Iron." *The Literary Digest*, March 17, 1934.
11 Science & Industry. "Chasing Meteorites is a Career." *Life* (September 6, 1937): 46-50.
12 Pyle, Ernie. "Dr. Nininger Will Quit Chasing Meteors When One Hits His Head." *Rocky Mountain News*, July 18, 1941.
13 "Ford-TWA Expedition Explores Scenic Wonder." *Ford News*, January 1942.

14 Wikipedia contributors, "Strange as It Seems," Wikipedia, https://en.wikipedia.org/w/index.php?title=Strange_as_It_Seems&oldid=1102960551

15 Wikipedia contributors, "Ripley's Believe It or Not!," Wikipedia, https://en.wikipedia.org/w/index.php?title=Ripley%27s_Believe_It_or_Not!&oldid=1105698383

16 Fleming, Roscoe. "Miner of Meteorites." *Esquire Magazine*, May 1942

17 "Arizona's Newest Natural Wonder," *Arizona Highways*, May 1942.

18 Nininger, H.H. and Roscoe Fleming. "Meteorite!" *Collier's Magazine*, December 11, 1943.

19 Hilton, John. "Star Dust Hunter." *The Desert Magazine* (May 1947): 5-8.

20 Wikipedia contributors, "Liberty (general interest magazine)," Wikipedia, https://en.wikipedia.org/w/index.php?title=Liberty_(general_interest_magazine)&oldid=1085661364

21 Fleming, Roscoe. "Bombs from the Blue." *Liberty*, January 11, 1947.

22 H. H. Nininger Papers 1864-1991, Arizona State University. University Archives. Box 9, Folder 17.

23 Nesbitt, Sterling and William Parker. "The 1947 Society of Vertebrate Paleontology Triassic Field Trip, Then and Now, and Its Significance to Triassic Vertebrate Paleontology. *Mesa Southwest Museum Bulletin*, January 2005.

24 Clark, Neil M. "Star Chaser." *Saturday Evening Post.* April 17, 1948.

25 Nininger, H.H. "On the Trail of a Meteorite." *Desert Magazine.* March 1948: 12-15.

26 McKelvey, Nat. "On the Trail of a Meteorite." *Arizona Highways Magazine.* May 1948: 23-25.

27 "Meteorite Mystery." *Newsweek*. May 7, 1951: 54.

28 Hunt, Morton M. "The Man Who Chases Meteors." *Argosy*. June 1952: 71.

29 Wikipedia contributors, "Morton Hunt," Wikipedia, https://en.wikipedia.org/w/index.php?title=Morton_Hunt&oldid=1105440957

30 Megargee, Hermine. "The greatest natural laboratory in the world." *Point West Magazine*. June 1961.

31 Berland, Theodore. "Millionaire Miner of the Milky Way." *Saga Magazine*. December 1962.

32 Scher, Zeke. "How Harvey Nininger's world was created." *Denver Post Empire Magazine.* May 25, 1975.

33 Weaver, Kenneth F. "Meteorites: Invaders From Space," *National Geographic,* September 1986: 391-405.

34 Trefil, James. "Stop to consider the stones that fall from the sky," *Smithsonian,* September 1989: 81-93.

Chapter 2

1 Nininger, H. H. *Find a Falling Star.* Paul S. Eriksson, Inc., New York. 1972. p 4. Unless otherwise noted, quotes in this chapter are from this source.
2 Farrington, Oliver C. 1915. "Catalogue of the meteorites of North America, to January 1, 1909." National Academy of Sciences. Volume XIII,Washington, D.C. p. 11.
3 Castillo, Antonio del. Catalogue descriptif des météorites (fers et pierres météoriques) du Mexique: avec l'indication des localités dans lesquelles ces météorites sont tombés ou ont été de-couverts. Paris: Imprimerie L. Ouin, 1889.
4 Nininger H.H. 1943: The Eaton Colorado meteorite: introducing a new type.." *Popular Astronomy* 51: 273-280.
5 Buseck, P. R., Holdsworth, E. F., & Scott, G. R. "The Eaton copper meteorite - brass from space?' *Meteoritics* 4: 267.
6 https://drbarakataly.wordpress.com/2015/05/23/the-problem-of-eaton-meteorite-origin-a-review/
7 Foote, W.M. "Preliminary note on the shower of meteoric stones near Holbrook, Navajo county, Arizona, July 19th, 1912, including a reference to the Perseid swarm of meteors visible from July 11th to August 22nd." *American Journal of Science* 34, Ser. IV, 1912: 437-456.
8 Reeds. Chester A. "Comets, Meteors and Meteorites." *Guide Leaflet Series No. 77.* The American Museum of Natural History, New York, 1933.
9 Nininger, H.H. 1932. "The Beardsley Meteorite." *American Mineralogist* 17(12): 563–566.
10 Nininger, H. H. "A Meteorite Fall on August 10, 1932, near Archie, Case County, Missouri." *Popular Astronomy* 44: 93.
11 Merrill, George P. 1924. "A New Find of Meteoric Stones Near Plainview, Hale County, Texas." *Proceedings U. S. National Museum* 52-2184: 419-422.
12 https://onlineonly.christies.com/s/deep-impact-lunar-rare-meteorites/complete-plainview-meteorite-h-h-nininger-provenance-23/8280.
13 Nininger, H. H., and J. D. Figgins. 1933. "The Excavation of a Meteorite Crater Near Haviland, Kiowa County, Kansas." *Proceedings of the Colorado Museum of Natural History* 12-3: 9-16.
14 Bibbins, A. B. 1926. "A small meteor crater in Texas." *Engineering and Mining Journal-Press* 121: 932.

15 Barringer Jr., D. M. 1928. "A New Meteor Crater." The Academy of Natural Sciences of Philadelphia, LXXX, 307-311.

16 Nininger, H. H. 1934. "The Odessa, Texas, Meteorite Crater." *Popular Astronomy* 42: 46-47.

17 Nininger, H. H. 1934. "A Gold-Bearing Stony Meteorite from Melrose, New Mexico." *American Mineralogist* 19(8): 370–374.

18 Hudson, D. T. 1972. "A Meteorite Discovery with Archaeological Importance from the Camp Verde Area, Central Arizona." *Plateau* 45(1): 41-43.

19 H. H. Nininger Papers 1864-1991, Arizona State University. University Archives. Box 1, Folder 41.

20 Tassin, W. 1902. "The Casa Grandes Meteorite." *Proceedings U. S. National Museum* 25: 69-74.

21 Heineman, Robert Emil S.and Lionel Francis Brady. 1929. "The Winona Meteorite." *American Journal of Science,* December 1929, s5-18 (108): 477-486.

22 Plotkin, Howard, Roy Clarke, Jr. 2012. "The Controversial History of the Goose Lake, California, Meteorite." *Earth Sciences History* 31(2): 229–246.

23 Ibid., 238.

24 Leonard, Frederick C. 1939 "The Goose Lake Siderite: California's largest known Meteorite." *Popular Astronomy* 47(6).

25 Watson, Fletcher G. 1941. *Between the Planets.* Harvard University Press, Cambridge, MA.

26 Nininger, Addie Delp and Harvey Harlow Nininger. 1950. *The Nininger Collection of meteorites: a catalog and a history.* American Meteorite Museum, Winslow.

27 Farrington, Oliver C. 1895. "Handbook and catalogue of the meteorite collection," Field Columbian Museum Geological Series, Volume I: 7.

28 Farrington, Oliver C. *Catalogue of the meteorites of North America, to January 1, 1909.* American Museum of Natural History. New York, 1925: 14.

29 H. H. Nininger Papers 1864-1991, Arizona State University. University Archives. Box 3, Folder 2.

Chapter 3

1 Reeds, Chester A. "Comets, Meteors, and Meteorites." *Guide Leaflet Series, No. 77.* The American Museum of Natural History. New York, 1933.

2 Hoyt, William Graves. 1983. "Meteor Crater: Historical Note on Nomenclature." *Meteoritics* 18(2).

3 Fairchild, H. L. 1907. "The Origin of Meteor Crater (Coon Butte), Arizona," *Bulletin of Geological Society of America* 19: 493-504.

4 Foote, A.E., 1892. "A new locality for meteoritic iron with a preliminary notice of the discovery of diamonds in the iron." *Proceedings of the American Association for the Advancement of Science* 40: 279-283.

5 *Ibid.*

6 https://ngmdb.usgs.gov/ht-bbin/tv_browse.pl?id=042f8ece3c0b323b18babde1086dda63

7 See note 3 above.

8 Merrill, George P. 1908. "The meteor crater of Canyon Diablo, Arizona; its history, origin and associated irons." *Smithsonian Institution Miscellaneous Collection.* 50: 461-498.

9 Dalron, N.H., et al. 1915. *Guidebook of the Western United States: Part C - The Santa Fe Route, with a side trip to Grand Canyon of the Colorado.* Bulletin 613. U.S. Geological Survey.

10 Chaffey, Pauline. 1938. "Meteor Mountain." *Arizona Highways,* September.

11 Decision Lists Nos.4607-4609, U.S. Department of Interior Board of Geological Names, July-September, 1946.

12 https://www.lpi.usra.edu/science/kring/epo_web/impact_cratering/enviropages/Barringer/barringerstartpage.html

13 https://www.nps.gov/subjects/nnlandmarks/site.htm?Site=BAME-AZ

14 Chladni, Ernst F. Friedrich. 1794. "On the origin of the iron masses found by Pallas and others similar to it, and on some natural phenomena associated with them." Johann Friedrich Hartknoch. Riga, Latvia.

15 Biot, J.-B. 1803. "Account of a journey that was made in the department of the Orne in order to confirm the reality of a meteor that was observed in L'Aigle on 26 April 1803. (in French). Paris, France: Baudouin.

16 Proctor, Richard Anthony. *The Moon: Her Motions, Aspect, Scenery, and Physical Condition*. London: Longmans. 1873.

17 Proctor, Richard Anthony. *The Moon: Her Motions, Aspect, Scenery, and Physical Condition*. Second Edition London: Longmans, Green, and Co. 1878.

18 Gilbert, Grove Karl. 1891. *Notebook 51: with Baker to Flagstaff, Coon Mountain, Canyon Diablo, mouth of San Francisco Wash.* Photocopy of notebook provided by U.S. Geological Service, Flagstaff, Arizona.

19 Gilbert, Grove Karl. "The Origin of Hypotheses, Illustrated by the Discussion of a Topographic Problem." *Science.* January 3, 1896.

20 Nininger, Harvey H. 1951. "A Résumé of Researches at the Arizona

Meteorite Crater." *The Scientific Monthly* 72 (2).

21 Much of the D.M. Barringer biographical material is take from: Barringer, Brandon. 1964 "Daniel Moreau Barringer (1860 - 1929) and His Crater." *Meteoritics* 2 (3) December.

22 Barringer, Brandon. Correspondence to H. H. Nininger dated April 10, 1956. H. H. Nininger Papers 1864-1991, Arizona State University. University Archives. Box 20, Folder 12.

23 Barringer, Daniel Moreau. 1906. "Coon Mountain and its crater." *Proceedings of the Academy of Natural Sciences of Philadelphia*. 57: 861-886.

24 See note 20 above, p. 186.

25 See note 21 above.

26 Tilghman, B. C. 1905, Coon Butte, Arizona, *Proceedings of the Academy of Natural Sciences of Philadelphia* 57: 887.

27 Merrill, G.P., 1908. "The meteor crater of Canyon Diablo, Arizona; its history, origin and associated irons." *Smithsonian Institution Miscellaneous Collection* 50: 461-498.

28 Barringer, Jr., D. M. 1927. "The Most Fascinating Spot on Earth," *Scientific American* 137: 52-54, 144-146, 244-246.

29 See Note 20 above, p. 196.

30 See Note 20 above, p. 190.

31 https://www.dollartimes.com

32 Nininger, H. H. Undated handwritten note entitled "The Ariz. Crater and Me." Nininger Files: Sedona Historical Society.

33 Nininger, H. H. 1956. *Arizona's Meteorite Crater. Past, present, future.* American Meteorite Museum, Sedona. p. 48.

34 Nininger, H. H. *Our Stone-Pelted Planet: A Book About Meteors and Meteorites*. Houghton Mifflin Company, Riverside Press Cambridge. 1932.

35 *New York Times,* 16 April 1933, Miscellaneous Brief Reviews section. "About Meteors."

36 Nininger, H.H. 1936. "A New Type of Nickel-Iron Meteorite, from the Vicinity of the Arizona Meteorite Crater." *Meteors and Meteorites*.

37 Heymann, D. 1964. "Origin of the Canyon Diablo No. 2 and No. 3 Meteorites." *Nature* 204: 819–820.

38 Reports "Grants of the American Philosophical Society." *Science*. 15 Dec 1939. 90(2346): 568-570.

39 Nininger, H. H. *A Comet Strikes the Earth*. American Meteorite Laboratory, Denver, Colorado. 1942

40 La Paz, Lincoln. 1947 "A Comet Strikes the Earth (Review)." *Popular Astronomy* 56: 158.

41 Nininger, H. H. *Chips from the Blasted Moon.* American Meteorite Museum. Winslow, Arizona. 1947.
42 Nininger, H. H. 1949. "A New Type of Magnetometer Survey at Barringer Meteorite Crater." *Popular Astronomy* 57:17.
43 H. H. Nininger Papers 1864-1991, Arizona State University. University Archives. Box 20. Folder 11.
44 *Ibid.*
45 H. H. Nininger Papers 1864-1991, Arizona State University. University Archives. Box 20, Folder 11.
46 Plotkin, Howard and Roy S. Clarke Jr. 2008. "Harvey Nininger's 1948 attempt to nationalize Meteor Crater." *Meteoritics & Planetary Science* 43 (10): 1741–1756.
47 Nininger, H. H. *Out of the Sky : An Introduction to Meteoritics.* University of Denver Press, Denver, Colorado. 1952.
48 Nininger, H. H. *Arizona's Meteorite Crater: Past, Present and Future.* American Meteorite Laboratory. Sedona, Arizona. 1956.
49 Rinehart, John S. 1957. "Special Issue Book Review." *Science* 26 (125) Issue 3252: 829.
50 H. H. Nininger Papers 1864-1991, Arizona State University. University Archives. Box 8, Folder 32.

Chapter 4

1 Most of the material relating to Harry Locke was obtained from the Harry Locke Collection archives at the Old Trails Museum/Winslow Historical Society.
2 Repp, Thomas A. *Route 66: The Romance of the West.* Mock Turtle Press. Lynnwood, Washington. 2002.
3 Smyth, Fred. Hayden Planetarium Sales Booth. Order Form to Harry Locke, 22 March 1937. Locke Archives: Old Trails Museum/Winslow Historical Society.
4 Federer, Helen. Hayden Planetarium Book Corner. Letter to Harry Locke, 25 September 1937. Locke Archives: Old Trails Museum/Winslow Historical Society.
5 Federer, Helen. Hayden Planetarium Book Corner. Letter to Harry Locke, 13 April 1938. Locke Archives: Winslow Historical Society.
6 H. H. Nininger Papers 1864-1991, Arizona State University. University Archives. Box 49, Folder 18.

Chapter 5

1 H. H. Nininger Papers 1864-1991, Arizona State University. University Archives. Box 30, Folder 8.

2 https://www.officialdata.org/us/inflation/on January 6, 2023.

3 Allan O. Kelly Geology Papers, 1872-1997. MS 014. Series. Folder. *Australia Down Under And Off To The Side Part 2.* Carlsbad History Collection, Carlsbad City Library.

4 Nininger, H. H. 1972. *Find A Falling Star.* Paul S. Ericksson, Inc. New York. p. 141.

5 Simpsonn, E. S. 1938. Some new and little known meteorites found in Western Australia. *Mineralogical Magazine* 25: 157–171.

6 Nininger, H. H. *Find A Falling Star.* Paul S. Ericksson, Inc. New York. 1972. p. 215.

7 Hamacher, D.W. and C. O'Neill. 2013. "The Discovery and History of the Dalgaranga Meteorite Crater, Western Australia." *Australian Journal of Earth Sciences*, 60: 637–646.

8 Nininger, H. H. 1958. Personal notebook on first Dalgaranga Crater visit. Arizona State University Buseck Center for Meteorite Studies.

9 See note 3 above, Figure 17 and Figure 18, unnumbered between pages 153 and 154.

10 Nininger, H. H. and Glenn Huss. 1960. "The unique meteorite crater at Dalgaranga, Western Australia." *Mineralogical Magazine* 32: 619–639.

11 *Ibid*, p. 627.

12 Mardock, Marsha. 2005. "Howard Madole." *Modern Phoenix.*

13 City of Sedona Historic Landmarks Inventory.

14 See note 5 above.

15 Nininger and Huss, p. 621.

16 Hamacher, p. 642

17 Discussion of the Arizona State University acquisition was obtained from https://meteorites.asu.edu/aboutus/history.

18 H. H. Nininger Papers 1864-1991, Arizona State University. University Archives. Box 46, Folder 13.

19 https://www.officialdata.org/us/inflation/ on January 6, 2023.

20 Nininger, H.H. *Ask A Question About Meteorites.* American Meteorite Laboratory, Denver, Colorado. 1961.

21 Millman, Peter M, 1962. "Review of Publications - Ask a Question About Meteorites." *Journal of the Royal Astronomical Society of Canada* 56: 233.

22 https://www.officialdata.org/us/inflation/on January 6, 2023.

23 Nininger, H. H. *Find A Falling Star.* Paul S. Ericksson, Inc. New York. 1972. p. 224.

24 Rigby, Elizabeth. "Mineral is Named After Sedona Man." *Arizona Daily Sun*, November 12, 1966.

25 *Meteorite Crater Study Kit*. American Meteorite Laboratory. Denver, Colorado. No copyright date shown.

26 Quote is from a page of the 1967 ceremonial program at McPherson Collage, bestowing honorary Sc.D to Nininger. Nininger Files: Sedona Historical Society.

27 Nininger, H. H. *Find a Falling Star*. Paul S. Eriksson, Inc., New York. 1972.

28 Gibson, Jr., Everett K. 1973. "Book Review: Find A Falling Star." *Meteoritics* 8: 186.

29 Cook, Allan F. 1975. "Book Review: Research under Difficulties: Find a Falling Star." *Journal for the History of Astronomy* 6(2): 140-141.

30 https://www.goodreads.com/book/show/701439.Find_a_Falling_Star.

31 https://www.google.com/books/edition/Find_a_Falling_Star/mN1-AAAAIAAJ?hl=en.

32 https://www.worthpoint.com/worthopedia/find-falling-star-nininger-meteorites-478179617.

33 https://aerolite.org/shop/historic-meteorites.

34 Letter from Pomona College President David Alexander dated April 5, 1976. Nininger Files: Sedona Historical Society.

35 https://minorplanetcenter.net/db_search/show_object?object_id=2421.

Made in the USA
Coppell, TX
19 February 2026

72419136R00112